Gerhard Steinebach • Subhrajit Guhathakurta
Hans Hagen
Editors

Visualizing Sustainable Planning

 Springer

Editors

Prof. Dr. Gerhard Steinebach
Universität Kaiserslautern
Fachbereich Informatik
67653 Kaiserslautern
Germany
steineb@rhrk.uni-kl.de

Prof. Dr. Hans Hagen
Universität Kaiserslautern
Fachbereich Informatik
67653 Kaiserslautern
Germany
hagen@informatik.uni-kl.de

Prof. Dr. Subhrajit Guhathakurta
Arizona State University
School of Planning & Landscape
Architecture
P.O. Box 872005
Tempe AZ 85287-2005
USA
subhro.guha@asu.eduo

ISBN 978-3-540-88202-2   e-ISBN 978-3-540-88203-9

X.media.publishing ISSN 1612-1449

Library of Congress Control Number: 2008943976

Cover design: KuenkelLopka, Heidelberg, Germany

Printed on acid-free paper

9 8 7 6 5 4 3 2 1

springer.com

Preface

This volume reflects the work of an interdisciplinary team of scholars engaged in the pursuit of sustainable planning through the development and use of innovative visualization technologies. Visualization is perceived as a comprehensive tool for analysis of large, complex, and unstructured data, as well as a means for effective communication. Given the core functions of planning in the Western democratic traditions, the visualization component is not only significant but also indispensable. Despite its essential role in planning, the planning literature on visualization technologies has not evolved beyond the use of Geographic Information Systems (GIS). In this volume we present the state of art in the rapidly growing field of visualization as it relates to problems in urban and regional planning.

Planning is about deliberating on, visioning, and guiding a community's future. It is also about collaborative problem solving, effective communication of options, persuasion, plan making, monitoring, and often political strategizing to enable communities to achieve their desired futures. Plans are the products of this complex process that is uniquely and fundamentally communicative in nature. Therefore the ability to communicate ideas and to critically evaluate and coalesce multiple voices are among the essential elements of planners' toolkit. Effective visualization tools provide the means of improving communication among the actors involved in planning and offer an analytical framework for examining how social actions lead to future quality of life impacts. In this volume we bring together a rich compendium of such visualization tools that have the potential to impact urban planning together with some examples of their use in practice.

The significance and timeliness of this volume rests on several developments in the literature and in the challenges facing cities. First, the unsustainability of many of our current paradigms of development has become evidently clear. We are entering into an era in which communities across the globe are strengthening their connections to the global flows of capital, goods, ideas, technologies, and values while at the same time facing serious dislocations in their traditional socioeconomic structures. We are now more aware of the "global village" not only in terms of the immediacy of our mutual exchanges but also because of the environmental and economic impacts that transcend geographic boundaries. While the impending scenarios of climate change impacts remind us about the integrated ecological system that we are a part of, the current discussions of global recession in the media alerts us to the occasional perils of the globalized economic system. The globally dispersed, intricately integrated, and hyper-complex socio-economic-ecological system is difficult to analyze, comprehend and communicate without effective visualization tools. Given that planners are at the frontlines in the effort to prepare as well as build resilience in the impacted communities, appropriate visualization tools are indispensable for effective planning.

Second, planners have largely been slow to incorporate the advances in visualization research emerging from other domains of inquiry. The research on visualizing 3-dimensional environments have now entered the mainstream of computer science with a number of highly cited articles. Other disciplines, such as graphic design, geography and cartography have also lead in the development of new forms of visualization and communication, both conceptually and technologically. In contrast, the literature on modeling and visualization in planning has relied heavily on geographic information systems (GIS) tools that continue to provide two-dimensional spatial maps in formats not significantly different from those of a decade ago. This is not to suggest that research on planning support systems and GIS have been stagnant. Integrated models of transportation-land use-environment have become more sophisticated and several operational models are currently in use. Regardless, visualization research in planning has not kept pace with these developments. This volume attempts to redress this gap in the planning literature.

Third, the specter of climate change has brought into sharper focus serious issues about our current empirical tools and their ability to address future risks and opportunities. Most empirical models are built upon well-behaved data that follow well-known distributions of probability. However, the significant history-defining events are often the most improbable "outliers". These significant occurrences, which are now labeled "Black Swan" events, are often rationalized after the fact. This process of post-hoc rationalization inevitably suffers from "hindsight bias". To speculate about such high impact and low probability events require imaginative thinking along many attributes that cannot be accomplished without engaging multiple perspectives. Visualizing futures offers an attractive option for both clarifying speculatively derived mental models of often improbable events and for communicating such models among a large group of stakeholders.

Finally, as we enter a world of information overload, it is becoming increasingly difficult to sift through the profusion of information and gather the useful parts for further analysis. Appropriate visualization techniques provide a means of focusing attention and communicating significant patterns that are often hidden within the information "noise". In this respect it is now an indispensable tool for analysis. The power of visualization in shaping attitudes privileges some ideas in ways that may also hide other critical truths. Hence it is also important to critically evaluate the assumptions that are embodied in visualizations and find ways to make visualization process a collaborative exploratory process. That is indeed one of the challenges of visualization when planning for sustainable development.

The researchers represented in this volume belong to an International Research Training Group (IRTG) that is coordinated by faculty from Technische Universitaet

Kaiserslautern and funded by the Deutsche Forschungsgemeinschaft. Over the past five years, the IRTG has graduated 47 doctoral students who were guided by 20 academic faculty members or research scientists. These graduate students, faculty, and other research personnel work in interdisciplinary teams working on a range of problems such as visualization in planning communication and participation, robotics for earthquake geology, visualization of higher order finite element methods, visualization of dynamic noise profiles, visualizing effects of climate change, among many others. Despite the wide range of topics in visualization, the diverse and interdisciplinary team has been able to forge an effective collaboration, which has enabled learning and communication across the domain boundaries. The chapters in this volume are intended to provide a flavor of the cross-pollination of ideas on visualization and their effectiveness in engaging planning problems.

A project of this scope and reach requires a well-knit and efficient organization with many committed individuals. Without their expertise, commitment, knowledge, and hard work, this volume perhaps would not have seen the light of day. First and foremost, we wish to thank the Deutsche Forschungsgemeinschaft for their support and encouragement, which made this IRTG a successful endeavor.

Gerhard Steinebach
Subhrajit Guhathakurta
Hans Hagen

Contents

Contributors

John Crittenden
Department of Civil and Environmental Engineering, Arizona State University,
Tempe, AZ 85287-5306, USA

Karthikeya Date
School of Architecture and Landscape Architecture, College of Design, Arizona
State University, Tempe, AZ 85287-1605, USA

Eduard Deines
Institute for Data Analysis and Visualization, University of California, Davis,
2144 Academic Surge, Davis, CA 95616, USA, edeines@ucdavis.edu

Achim Ebert
Technische Universität Kaiserslautern, Fachbereich Informatik, HCI & Visualization
Lab, Postfach 3049, 67653 Kaiserslautern, Germany, ebert@cs.uni-kl.de

Katja Einsfeld
Technische Universität Kaiserslautern, Fachbereich Informatik, HCI & Visualization
Lab, Postfach 3049, 67653 Kaiserslautern, Germany, einsfeld@informatik.uni-kl.de

Robin Ganser
Department of Planning, Oxford Brookes University, Headington Campus, Gipsy
Lane, Oxford, OX3 0BP, UK, ganser@brookes.ac.uk

Susanne Grossman-Clarke
Global Institute of Sustainability, Arizona State University, P.O. Box 872511,
Tempe, AZ 85287-2511, USA, sg.clarke@asu.edu

Subhrajit Guhathakurta
School of Planning, College of Design, Arizona State University, Tempe, AZ
85287-2005, USA

and

School of Planning and Global Institute of Sustainability, Arizona State
University, Tempe, AZ 85287-2005, USA, subhro@asu.edu

Hans Hagen
Technische Universität Kaiserslautern, Fachbereich Informatik, Postfach 3049.
67653 Kaiserslautern, Germany, hagen@informatik.uni-kl.de

Martin Hering-Bertram
Fraunhofer-Institut für, Techno- und Wirtschaftsmathematik, Fraunhofer-Platz 1,
67663 Kaiserslautern, Germany, martin.hering-bertram@itwm.fhg.de

Janet Holston
Herberger Center, College of Design, Arizona State University, Tempe, AZ
85287-1905, USA

Khanin Hutanuwatr
College of Design, Arizona State University, PO Box 872105, Tempe, AZ 85287-
2105, USA, khutanuw@asu.edu

Joonwon Joo
Multimodal Planning Division, Air Quality Policy Branch, Arizona Department
of Transportation, 206 S. 17th Ave. 320B, Phoenix, AZ 85007, USA, jjoo@azdot.
gov

Yoshi Kobayashi
School of Architecture and Landscape Architecture, College of Design, Arizona
State University, Tempe, AZ 85287-1605, USA

Goran Konjevod
School of Computing and Informatics, Arizona State University, Tempe, AZ
85287-8809, USA

Tim Lant
Decision Theater, Arizona State University, Tempe, AZ 85287-8409, USA

Jesus J. Lara
Austin E. Knowlton School of Architecture, The Ohio State University, 291
Knowlton Hall, 275 West Woodruff Avenue, Columbus, OH 43210-1138, USA,
lara.13@osu.edu

Ke Li
Department of Civil and Environmental Engineering, Arizona State University,
Tempe, AZ 85287-5306, USA

Daryl A. Lloyd
Centre for Advanced Spatial Analysis, University College London, 1-19 Torrington
Place, London WC1E 7HB UK, daryl.lloyd@dunelm.org.uk

Frank Michel
Deutsches Forschungszentrum für Künstliche Intelligenz, Trippstadter Str. 122,
67663 Kaiserslautern, Germany, frank.michel@dfki.de

Ariane Middel
Decision Center for a Desert City, Global Institute of Sustainability, Ari-zona
State University, P.O. Box 878209, Tempe AZ 85287-8209, USA, ariane.middel@
asu.edu

Jan Mohring
Fraunhofer-Institut für Techno- und Wirtschaftsmathematik, Fraunhofer-Platz 1,
67663 Kaiserslautern, Germany

Filiz Ozel
College of Design, Arizona State University, PO Box 871905, Tempe, AZ 85287-
1905 USA, ozel@asu.edu

Robert Pahle
College of Design, Arizona State University, PO Box 871905, Tempe, AZ 85287-
1905 USA, robert.pahle@asu.edu

Mookesh Patel
Department of Visual Communication Design, College of Design, Arizona State
University, Tempe, AZ 85287-2105, USA

Ray Quay
College of Design, Arizona State University, PO Box 872105, Tempe, AZ 85287-
2105, USA, ray.quay@asu.edu

Martin Rumberg
Technische Universität Kaiserslautern, Lehrstuhl Stadtplanung, Pfaffenbergstr.95,
67663 Kaiserslautern, Germany, rumberg@rhrk.uni-kl.de

Theo G. Schmitt
Technische Universität Kaiserslautern, FG Siedlungswasserwirtschaft, Postfach
3049, 67653 Kaiserslautern, Germany, tschmitt@rhrk.uni-kl.de

Gerhard Steinebach
Technische Universität Kaiserslautern, Lehrstuhl Stadtplanung, Pfaffenbergstr.95,
67663 Kaiserslautern, Germany, steineb@rhrk.uni-kl.de

Martin Thomas
Technische Universität Kaiserslautern, FG Siedlungswasserwirtschaft, Postfach
3049, 67653 Kaiserslautern, Germany, mthomas@rhrk.uni-kl.de

Ingo Wietzel
Technische Universität Kaiserslautern, Lehrstuhl Stadtplanung, Pfaffenbergstr.95,
67663 Kaiserslautern, Germany, wietzel@rhrk.uni-kl.de

Joseph A. Zehnder
Department of Atmospheric Sciences, Creighton University, Hixson-Lied #541,
2500 California Plaza, Omaha, NE 68178-0002, USA, zehnder@creighton.edu

I
Planning

Planning Sustainable Living

Gerhard Steinebach

Introduction

The planning of future spatial development, in line with the social, environmental and economic dimensions of sustainability is one of the core tasks of spatial planning, which can be described as the development of cities, villages and countryside into a liveable environment, which respects people's needs [2: 897]. In doing so, the different spatial requirements [2: 894] which can complement, overlap and compete with each other have to be brought to a balance.

Based on the characterization of the development of urban planning in Germany since the 1960s, the overall concept of sustainability in spatial planning should be clarified by following the principle of sustainability through different time periods. The successful implementation of this means the realization of sustainable development on all spatial levels, including urban planning itself [2: 679]. The different planning levels interact between strategic planning and its implementation for achieving local objectives and measures. A careful and economic use of the ground is an important task of sustainable settlement development. Two examples from Germany and the United States should provide a practical context to the much more complex topic "Planning Sustainable Living". Both cases show planning for future settlement development under special conditions. The principles of sustainability are thereby taken into account under specific social, economic and environmental conditions.

1 Development of Urban Planning in Germany

Besides the arrangement and formation of space urban planning also comprises the coordination of all relevant actors in the course of planning. The topics represent a multitude of different issues from social, ecological or economic questions up to the

G. Steinebach
Technische Universität Kaiserslautern, Lehrstuhl Stadtplanung, Pfaffenbergstr.95,
67663, Kaiserslautern, Germany, E-mail: steineb@rhrk.uni-kl.de

G. Steinebach et al. (eds.), *Visualizing Sustainable Planning*,
© Springer-Verlag Berlin Heidelberg 2009

process of steering, moderation and mediation [3]. In this respect, urban planning represents a combination of different disciplines within one task.

In its development urban planning had to deal with the social, ecological and economic trends: Economic globalization, social individualization, liberalization and in Europe demographic change, to name a few. In consideration of urban planning since the 1960s some strings are visible that characterize the single eras. The objectives, which underlie the planning ideas of the respective developments, have experienced significant changes [3: 8].

Fig. 1 Märkisches Viertel, Berlin [20: 254]

For many citizens of the Federal Republic of Germany, the sixties of the last century was a time when they enjoyed a prosperous economy, broad mass-motorization, growing social wealth and population. The rising spatial and social mobility led to a migration from the polluted inner city to new residential areas in the outskirts. Despite the model of this era – technically everything is possible – an oversized housing scheme with high residential density and housing space was instituted. These areas were additional to the city centre and extended the supply to the suburban area. Ideally, the city planners should have completed and extended the core cities, and created urbanity by increasing density to serve as a counterweight to the uninhibited sprawl of the single-family houses in the outskirts [3: 53].

The "Märkische Viertel", a city quarter in the northern part of Berlin, obeyed and followed this maxim: Built from 1963 to 1974 and planned for 17,000 housing spaces and 40,000 inhabitants, it represents an expansion of room for housing for the metropolis. The facilities of public supply and education in the centre of the quarter underline its self-sufficient character.

This development was unsoldered in the following decade by sanitation and the large-area demolition of the historic city centres in the sense of a total zone-clearance. The beginning of the urban renewal was represented by a large-scale redevelopment. Fortunately, this was removed by selective clearance that gave attention to the traditional values of a constructional and social structure. The renewal and appreciation of existing buildings were accompanied by a social plan for the affected citizens. The change in understanding of planning from total demolition and replacement by new buildings to a method that focused on single objects and their importance in the social and constructional context was also a consequence of particular involvements of the citizens in the planning process. Since that time, the conservation of historic building stock has been quite important for urban policy. The simultaneously installed social planning reduced negative effects on the involved inhabitants. It also detected possible conflicts in advance and avoided them. From then onwards, sanitation described no longer the preservation of buildings but also considered the elimination of social grievances like cultural heritage, traditional architecture or the protection of socially stable neighbourhoods. At the same time the interest and request of the citizens in the emancipated democratic society was promoted. The sanitation of the "Ostertorviertel" in Bremen is a good example for the development process in sanitation and the shift from clearance to preservation.

Fig. 2 Ostertorviertel Bremen – planned track [9: 87]

The interest of citizens in their surroundings intensified in the 80's of the last century and focused itself in the field of ecology. Since then ecology planning and urban ecology as a part of it were established as a field of activity. The preservation of the natural basis of life has moved strongly to the centre of planning, because a newly discovered social consciousness has demanded the protection and care of the environment. Consequently, every planning project is judged for its compatibility with the surroundings. Thereby, an important request has been the avoidance and the decrement of unhealthy effects through emissions: The close interlocking and the aerial overlays of multiple use requirements on a restricted area brings noise screening to the fore. Within the boundaries of precaution planning it was applied to identify potential sources of loads of bordering areas and their effects Urban planning could therefore react, minimise restrictions restrictions as well as revise measures. The instrument of noise abatement by way of example of a city quarter in Kaiserslautern clarifies the development phase in urban planning as successful.

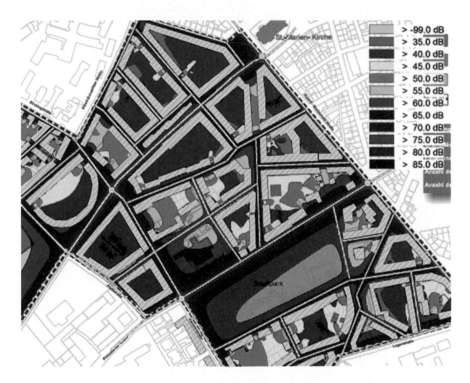

Fig. 3 Presentation of the results of a strategic noise mapping, Kaiserslautern [24]

The field of activity in urban planning in the 90's of the last century was affected by a strengthened economical view. As a result of the German reunification in 1990, planning didn't correspond any longer on the creation and expostulation of an area offering, but rather oriented itself after the particular demand. The involved project orientation and the performance in precise plans and infrastructure developments resulted on the one hand in a higher efficiency from the exploiture of available finance and budget resources. On the other hand a long-term integration of private engagement was accomplished in urban development. The Leipzig Train Station represents an example of this. At the beginning of the 90's of the last century the Leipzig Train Station was in extreme need of rehabilitation. Today it represents an architectural and urban creative climax of Leipzig. The Station with its 24 departure platforms is deemed to be on of the biggest head train stations in Europe. It is the hub of the city traffic as well as the central European local and national traffic with a daily charge of 150,000 commuters, passengers and visitors [37]. Within the phase of reconstruction the main building was enhanced by an atrium, which sheltered the "Hauptbahnhof-Promendaden" with a multitude of different supply and service proposals since 1997.

Fig. 4 Main Train Station, Leipzig, interior view [35]

Paying attention to sustainability in its social, ecologic and economic dimensions represents the combination of the already listed tasks in urban planning since the turn of the millennium. Sustainable urban development including foresighted land-use management can help to coordinate the demands of different duties to keep options of planning alive long-term (see Chapter "Visualizing Planning for a Sustainable World). The social aspect describes basically consequences of decreasing population in industrial nations for economy and society (e.g. in Germany the shift from 82.5 million inhabitants in 2002 to predicted 69 to 74 million inhabitants in 2050) [26]. The ecological aspect means taking care not to run out of resources. The change of the economic structure and its consequences on existing areas is a content of the economical point of view. Residential analysis and an urban settlement structure concept serve as examples for the idea of sustainable urban development. Analysis of the urban settlement and evaluation of future locations as well as larger related areas for potential residential use is the central focus of interest. Furthermore the urban settlement structure intent of Wiesbaden is in focus [28]. A multi-level analysis and evaluation system with elements of classic residential building land analysis and the English "urban capacity study" led to evaluating locations by quantitative as well as qualitative indicators. The aim of this procedure was to find areas that make best quality habitation in a first-class neighbourhood possible [28]. To support a sustainable social, ecological and economical development, the inner city development was in focus because different aims could be accomplished at the same time.

Fig. 5 Triangle of sustainability [own design]

As an example of respect for the sustainability concept of urban development, the residential site analysis should be related to a settlement structure concept for the city of Wiesbaden The analysis of the settlement area, the identification and evaluation of future locations and specific larger contiguous residential areas, and the development potentials of planning ideas for the settlement structure of Wiesbaden are in the spotlight [28].

2 The "Leitbild" of Sustainability

The development of the "Leitbild" of sustainability was primarily supported through international cooperation as well as intense economic and social exchange.

The tasks resulting out of the changing social, ecological, and economic conditions on a global scale make coordinated and cooperative action plans inevitable [11]. Examples of these developments are the ongoing globalisation and internationalization, the influence of information and communication-technologies, the ageing and shrinking of the society in greater parts of Europe, and the challenges of the global climate change.

In this context the discussion on consideration and implementation of sustainable development has to be scheduled both on the strategic and international level and has to be pursued down to the local level in the specific nation states [27]. This continuous process is affected by several milestones in the (inter)national discussion on sustainability on different chronological and hierarchic levels. Some of the most significant milestones will be exemplified in the following discussions to clarify this development.

2.1 Milestones of Sustainability

Sustainable development as a political objective was significantly influenced through the report "Our Common future", published by the UN World Commission on Environment and Development (WCED) in 1987. By definition sustainability describes a "development that meets the needs of the present without compromising the ability of future generations to meet their own needs" [31: 24]. This approach includes on the one hand the responsibility for the next generations and the responsible use of resources and on the other hand the responsibility within one generation-like the responsibility of the industrial countries towards the Third World countries [30].

To transform the named tasks of the report "Our Common Future" into a binding form the United Nations Conference on Environment and Development took place in Rio de Janeiro in 1992. With 178 participating countries the overall goal of the conference was the enhancement of the objective of sustainability into legal guidelines for its ecologic, economic and social dimensions. These guidelines were taken together in the Rio Declaration on Environment and Development that consists of 27 principles of sustainable development. There the link between global developments with the resulting problems and ecological objectives was in the focus. They were enhanced in various documents with statements on climate protection and protection of species as well as on implementation strategies on a national level [38]. The action program of the international community of states for the 21st century "Agenda 21" focused on the implementation of the goals of sustainable development on the national levels through plans measures and instruments. Chapter 28 of the Agenda concentrates on the reference between the global aspects of sustainable development and the local level. The Local Agenda 21 calls upon the local authorities to participate in the process of sustainable development and to implement their actions into a global context [30], [38].

Based on this in 2002 a follow-up conference was held in Johannesburg, South Africa, where the matters of the Rio conference were deepened and diversified. In this context issues like globalization, alleviation of poverty, resource conservation and efficient use of resources through sustainable energy policy and water management were broached [43]. The Bali "Climate Summit" which took place in December 2007 was the last meeting of the international community of states. It questioned the aspects of sustainability in view of the global climate change.

In the course of further development of the worldwide discussion on sustainability various focal points emerged which were issued in the 1996 Habitat II-conference of the United Nations in Istanbul, Turkey, as well as in the Urban 21-conference on the future of the cities, which took place in Berlin, Germany, in 2000. There the development of the cities under the various aspects of the different dimensions of sustainability was examined and transferred onto settlements and their structures. The major challenge of merging the environmental conditions (Rio conference, Agenda 21, Bali Climate Summit) and urban development (Istanbul conference, Urban 21) characterizes the complex tasks on a local level that include the equalization of the different objectives of the various stakeholders as well as the equalization of the economic, ecologic and social interests.

In Germany the idea of sustainable Development was enhanced through the involvement in the global discussion. Under the slogan "Think Global – Act Local" the "Local Agenda 21" is not only focused on the international responsibility of Germany for the successful implementation of sustainable development but also integrates all relevant stakeholders on the local level into the process.

Based on this the German "National Sustainability Strategy" shaped out the principle of sustainable development in the purpose of the Rio conference as a political objective for the modernization of policy and society. So the strategic implementation of sustainability is interconnected through indicator-based objectives and measures with statements on management, realization and continuous evaluation [11].

One of these objectives amongst others is the reduction of land-use in the context of settlement development through the consideration of the "30-ha-target" issued by the German Government or the "3:1-target" as an outcome of the recent research project "Future Cities". For one it's about the reduction of the daily land use for settlement purposes in Germany from 118 hectares per diem in 2005 to 30 hectares per diem in 2020 [36]. For another the use of previously developed land in the core areas of the cities and the use of undeveloped land is set at a ratio of 3:1 with the goal to focus consequently on infill development [13].

2.2 Sustainable Urban Development

The meaning of the "Leitbild" of sustainable development is amongst others reflected in its influence on almost all aspects of life. One of the most significant aspects affected through the idea of sustainability can be seen in urban development.

Urban development depends heavily on the co-operation between politics, economy and society to fulfil the goals of sustainability. Out of the responsible dealing with the available resources an agreement between the interests of the present and future city dwellers is possible that supports settlement development in terms of sustainability. The urban functions are determined by the above-named impacts of economical, technological and social changes.

Current trends of urban development are significantly affected through demographic and social factors like ageing, effects of migration, and moving and diversification of lifestyles. Furthermore, economic factors like the decreasing importance of traditional site conditions, the growing part of brown fields or the ongoing reduction of jobs affect the settlement development. Therefore cities and regions are in a state of constant change and urban development is the result of various social and economical land-use requirements [11].

Cities and Communities have to contribute their part to the implementation of the strategic goals and preliminary considerations of sustainable development on the operative level through clearly defined measures. On the local level the spatial, economic and social circumstances, the intentions respecting possibilities of development as well as the involved stakeholders have to be considered case-by-case [34]. The various circumstances are also reflected in the different tasks that rise in the context of sustainable urban development like the efficient planning and spatial distribution of the functions in growing regions, and the guarantee of minimum standards, or the reorganization of usages in shrinking regions.

With the following examples it shall be pointed out how the idea of sustainability can be implemented into settlement development. Within the "Settlement Development Wiesabden 2020" possible housing areas in the German city of Wiesbaden were identified and in a next step drafts for a possible development on these sites were designed. All this happened under the terms of sustainability.

The case study "Sustainable Living Phoenix" showcases the fictional conversion of a previously developed inner city area in Phoenix, USA, into a residential area under the aspects of sustainability.

3 Best-Practice-Example: Settlement Development Wiesbaden 2020

3.1 Municipal Framework – Introduction and Context

The city of Wiesbaden has nearly 2000 years of history with its early beginnings as a Roman sentry. Today, Wiesbaden is the state capital of Hessen, one of the 15 Federal States of Germany.

Wiesbaden has a major focus on the supply of services and retail, as an administrative centre, and an important city for federal and state authorities Manufacturing and industry are based in Wiesbaden, too.

Wiesbaden is located in a very important economic region in the western part of Germany and is part of the urban agglomeration Frankfurt/Rhine-Main. The metropolitan region Rhine-Main is a polycentric, compressed space with strong inner-economic, political and functional connections. With its bank centre of Frankfurt am Main and the Rhine-Main International Airport hub for goods services and flows of finance and information, it is also one of the most economically important European metropolitan regions in Germany.

Wiesbaden is on the eastern bank of the Rhine river, across the capital of the federal state Rhineland-Palatinate, Mainz. As the second largest city in the federal state of Hessen, Wiesbaden extends over an area of 204 km². Within the city limits live 274,964 inhabitants [42], out of whom 12,000 are American soldiers stationed in the Army Airfield in Wiesbaden-Erbenheim. Combining the metropolitan area of Wiesbaden with the surrounding communities and counties, about 570,000 inhabitants live here. The whole surrounding of the Rhine-Main metropolitan region has about 5 million people [40].

Due to its position in the Rhine-Main metropolitan region, Wiesbaden is an important work site, and by its convenient location and high quality of life a preferred residence in the Rhine-Main area. Therefore Wiesbaden is ranked in nationwide comparisons in the leading group of cities with the highest rents and real estate prices.

Fig. 6 overview Germany [own design]

Against the background of continuing growth pressures in the Rhine-Main region, the city cannot renounce the designation of additional construction. The federal state capital Wiesbaden expects an increase in the residential population from about 7,000 inhabitants to a total of almost 280,000 inhabitants in 2020 [25: 2].

This corresponds with the projected population growth throughout the district of Darmstadt of 3,856,800 inhabitants with a total increase of population of approximately 2.5% compared to the year 2002 [15: 4, 32].

Due to the projected demographic and urbane-structural development, it is assumed that the state capital will have a housing floor space of about 15,000 to 20,000 accommodation units up to the year 2020. That development runs parallel with an increase in the number of households in the same dimension [25].

According to the requirements of sustainable development for the area of Wiesbaden the planning task was to create a residential area analysis and housing structure concept as a basis for the continuation of the land-use plan [4].

Fig. 7 Metropolitan Region Rhine-Main [own design]

The analysis of the housing area, the identification and evaluation of future residential sites, and the potentials for development of planning ideas for the future housing structure in Wiesbaden were among the aims of this task. The urban and landscape environment including existing qualities and the situation of the social infrastructure were to be considered and respected in particular.

For the potential housing sites, proposals would have to be made furthermore for the planning development and the action of planning instruments. The aim to maximizing the saving of area and supporting a sustainable settlement structure development had to be pursued. The socially responsible, long-range supply of housing areas needed to be ensured for broad levels of the population, taking into account the particular spatial and structural circumstances of Wiesbaden.

In the end, a multi-level analysis and accordingly evaluation methods were developed which interlink methodical elements of classic residential land analysis with the concept of a so-called "Urban Capacity Study", which is standardized in the English planning system.

In order to properly express both quantitative and qualitative results, all phases of analysis were described in detail and evaluated.

In the first phase, the actual relevant restrictions based on a set of indicators were identified and overlaid.

Restriction criteria were for example: the conservation of water protection areas, unfavourable topography or spatial or functional conflicts from different land uses such as supply and disposal sites, etc. [fig. 12: 22].

The result of the extensive restriction analysis was a "black-and-white illustration" of areas that are filled up with restrictions and therefore have been segregated, as well as such areas that contain potential housing sites.

In the following analysis-phase, those locations were examined where conflicts like problematic land uses, compensatory land, or even building-protect areas, and an unsatisfying connection to the public transport system could be expected.

As a result of the conflict analysis potential areas which showed the accumulation of conflicts were either definitively scrapped or assigned to a so-called "reserve pool", including areas which were considered to perform an immediate non-residential-development under the actual planning-horizon, on the basis of the expected conflicts, but would be worth considering for a settlement development in the long-run.

As a result, 20 potential sites were identified as being suitable for residential use.

In contrast to the previous analysis phase, the following territorial suitability analysis was a direct comparison of all grounds facilitated by a multidimensional approach to the assessment of suitability criteria: existing structures, interference intensity, noise and air pollution as well as the situation in the urban area, efficiency of the road network, facilities and quality of social infrastructure, or the ownership [fig. 12: 22].

A "ranking" of the overall areas came as a result of this step.

Part of the suitability assessment was the development of a multi-density model, which aimed at using the resources "land" and "space".

In this context, it was particularly important to realize quality living in a high-quality living surrounding. Therefore the orientation of the existing municipal structures and the further development of these structures in urban design was a goal of sustainable urban development.

For six selected potential areas detailed test designs were created with housing and building typologies in consideration of the mentioned criteria and a reasonable urban density in an appropriate mixture. These would prove the feasibility of the area-related sustainability requirements, taking the existing environment and the specific location in the urban surroundings into account.

Subsequently, the transferability of the design results was checked for similar potential areas. Consequently it was observed that coverage of the required housing units would be achieved in a realization similar to the test designs. Thus, different locations and area-planning alternatives could be shown by the settlement structure concept, representing the impact on an updated land-use plan. This was the technical basis for a discussion of various political "guiding principles" for the purposes of superior conceptual urban development principles. Active participation of the population of Wiesbaden was a necessary step for a sustainable residential land development.

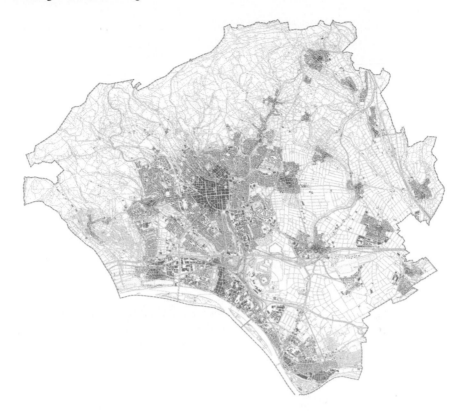

Fig. 8 District of Wiesbaden [digital map, land surveying office Wiesbaden]

3.2   *Methodology*

Methodically the process covered a combined sequential and interactive housing site analysis including the necessary on-site examination as well as the so-called "Urban Housing Capacity Study" [16]. The housing site analysis and the resulting settlement development concept are subdivided into various steps, which will be specified in the following.

3.2.1   Area-Wide Restriction Analysis with One-Dimensional Evaluation Approach

The definition of all relevant criteria, in particular the restrictives [fig. 12: 22] led to the elimination of areas which were not suitable for the realization of new housing sites or those which were already stated as legal housing areas and therefore weren't in the focus of the analysis.

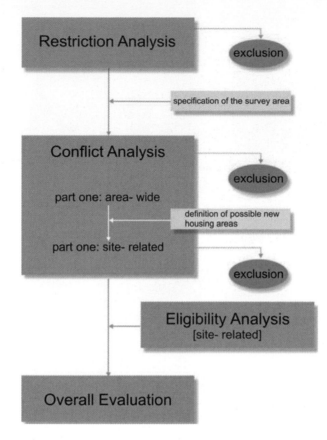

Fig. 9 Procedure housing site analysis [28: 23]

The analysis that was carried out was both area-related and separately considered for every single restriction. The results were documented in a map covering the whole area of Wiesbaden.

A further step included the overlay of the identified restrictive areas and the possible areas for future housing development that were not affected by restrictions. This resulted in an area-wide map in terms of an "aerial black-and-white picture".

Based on this area-related presentation of the restrictions, the exact boundaries of the possible new housing areas were defined.

3.2.2 Determination of Possible Housing Areas

After considering the existing planning, the settlement development of the past as well as the professional knowledge of the municipality, 15 possible new housing

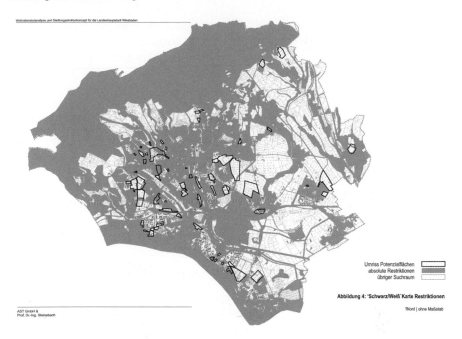

Fig. 10 "Black-and-White map" of restrictions [28: 51]

areas were additionally included into the examination. Regarding the intended reduction of land-use the extended focus was mainly on the inner-city areas.

A further survey of the possible areas took place in a multi-level process. This was done first through analyzing aerial photographs of the identified areas, which weren't affected through restrictions. As a result this analysis provided fallow or under-used areas that are capable of new housing development.

According to the particular demands the identified areas were verified through desktop-studies, on-site examinations, and expert-interviews. Thereby those areas were included which were likely to become housing areas through conversion in the near future.

All the identified areas for possible housing development were environmentally analyzed in the following ways.

3.2.3 Conflict Analysis

To avoid conflict of interests and to optimize the search for future housing-areas all the identified areas did undergo an environmental analysis.

There the focus was set on the existing land use patterns as well as the urbanistic, ecological and judicial backgrounds. Related to the environmental protective goods the analysis took place area-wide. In consideration of the intention of the study, the available data on the part of the city, and to avoid disproportional inquiry efforts some criteria were surveyed only on a site-related basis (e.g. noise emissions).

Criteria that could lead to exclusion or postponement of the identified areas were agreed upon with the city of Wiesbaden in the run-up to the analysis. The evaluation of the single areas didn't happen through simple summation of the appearing conflicts because they had to be valued very differently in their relevance amongst each other as well as in their impact on the particular site. Regarding this the evaluation took place strictly through single valuation under consideration of the particular site-characteristics.

In the result the analysis was differentiated into:

- Areas that were excluded because of the identified conflicts that made it inadvisable to develop them for housing purposes.
- Areas that were postponed due to the analysis that rated them available for housing development only in the long term. Those areas were added to a so-called "standby pool" of areas.
- Areas that could be developed for housing purposes in the given time horizon.

Another result of the analysis was the preliminary estimation of the overall size of the potential housing areas that consider the existing conflicts. Areas that were expected to be used for housing development within the given time horizon underwent an eligibility analysis related to the particular areas.

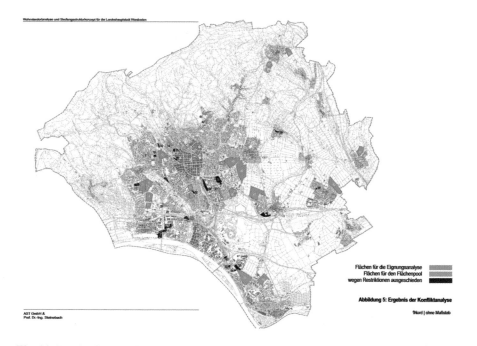

Fig. 11 Result of the conflict analysis [28: 87]

3.2.4 Site-Related Eligibility Analysis and Drafts

This eligibility analysis of the potential areas that could be developed for housing purposes within the time horizon of 2020 was used to compile a ranking for those areas. In this ranking the creative, ecological, social, economic, and infrastructural criteria, as well as the existing ownership structure were considered in particular.

As a result of this the evaluation of the specific eligibility of the particular area in terms of a "site aptitude" was possible.

The eligibility analysis consisted of the following four elements:

- Evaluation of the potential areas on the basis of eligibility criteria with a multi-dimensional approach.
- Formation of area categories respective to site categories.
- Designing of drafts on selected areas.
- Transferability checks of the draft results on areas of the same category.

The final result of the eligibility analysis was the rating of the possible development on the particular areas based both on quantitative and qualitative criteria. This rating of the areas did undergo a recapitulatory consideration and overall evaluation, which allowed the comparison between the particular areas.

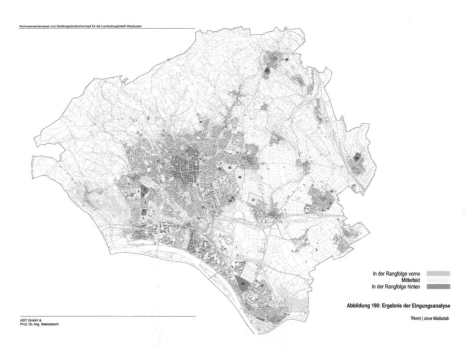

Fig. 12 Result of the eligibility analysis [28: 402]

In addition the parameters of the drafts were defined primarily out of the results of the conflict and eligibility analysis, the predetermined requirements, and the strived urban density. In addition to this the urban environment as well as possible approaches for traffic development had to be considered.

Subject	Restriction Analysis	Conflict Analysis	Eligibility Analysis
Land use / Plannings / Urban development	- existing / intended areas for housing and mixed- use development - areas of supply/ disposal - areas of primary production	- present land use - conflicting urban plannings - areas for securing the supply of raw materials - areas for agriculture	- present urban plannings - existing land use - surrounding land use - existing development structures - surrounding development structures
Protective gods/ Sanctuaries	- nature protection areas - proprietary biotopes - natural monuments - buffer areas - water protection areas	- landscape conservation area - suggested buffer areas - agricultural land of high ecological level - water protection areas - monument protection	- intensity of intervention
Influence (immissions and environment) on areas	- noise - air pollutants	- noise - air pollutants	- noise - air pollutants
Safety	- flood areas - areas of landslide	- protective areas	
Area and location	- area < 1 hectare - unpropitious topography/ orientation - missing affiliation to existing settlement areas		- size and expandability - location - inner- city - district - periphery - incoherent
Accessibility of public transport		- missing transport service / poor quality	- connection - planning
Road network			- capability - quality of roads and intersections
Supply			- accessibility within walking distance - primary care
Infrastructure			- existing social infrastructure
Lawn and open space			- quality and accessibility within walking distance
Estate			- number of owners - ownership structures
Economic efficiency			- brownfields/ contaminated sides - archaeological monuments

Fig. 13 Overview – analysis criteria [28: 27]

3.2.5 Recapitulatory Consideration and Evaluation

Both consideration and evaluation of the analyzed areas resulted out of the holistic look at the possible conflicts, the eligibility criteria and particularly the experimental drafts.

In the course of this a sequential approach for the development of the areas was worked out which included a ranking and mapping of the areas related to their "site aptitude", their chronological availability, the urbanistic priorities, as well as the needs of the fulfilment of demand.

Based on the results of the analysis and the guidelines for the drafts a superordinated guiding theme for the future settlement development in the city of Wiesbaden was formulated. Through this the background of a "Leitbild" shall be created which can be resolved consensually through administration, politics and civil society.

3.3   Concept of Settlement Structure

Besides finding and valuing future potential land for housing, this examination deals with planning the future development of Wiesbaden's structure of housing.

Determined demands, availability of areas, connection of locations (at the given period of time of planning), and a sustainable urban development in the center were considered in this concept. Therefore, a concept for an economic plan using housing development was worked out based on the idea of a sustainable urban development as a basic objective for Wiesbaden. The concept established a connection to the National Sustainability Strategy [8], which defines a quantified aim regarding the reduction of future land use for settlement and traffic. This aim refers to a reduction of additional land use down to 30 ha a day in 2020 ("30 ha-Ziel") [8: 99f.]. At the same time, a connection to another aim was established: A proportion of 3:1 concerning the inner and the outer development of cities [8: 296]. Several basic objectives were discussed for the concept of structure of housing, among them the "Compact City", the "Walkable City" and the "Polycentrality".

Therefore, considering the state of science, which postulates several suitable basic objectives for different levels of scale and specific locations instead of one structure of housing, a one-sided focus on compactness and density was not recommended but a detailed consideration of structure of settlement. This included different levels of mixed use, sizes of block structures, use and mix of type of buildings, and other parameters of settlements and housing. Over and above that, the connection with the city in general and the urban region were considered. This included the development of new urban junctions, the aimed process of partial spaces becoming denser, the precedence of inner over outer development, revitalization of fallow areas, formation of clusters and polycentrality, as well as consideration and further development of city and landscape appearance.

To keep the development of housing with a land use as economic as possible large connected areas were delimited in order to accommodate the medium-term and long-term demands of areas of housing. There were no large areas of former used areas available.

For a sustainable and economical development of housing, the monitoring of the known and available inner city areas (especially fallow areas, brown fields and empty sites) was mentioned. Within the scope of the examination, the release and development of the identified areas were to be managed.

The following tasks were determined:

- The required "critical mass" for utilities and services would be guaranteed.
- Development would be focused upon large locations in order to minimize a city's unnecessary spread over an area in several smaller locations.
- A dispersed development would be avoided.
- Sufficient areas for housing would be available during the whole time of planning.

Therefore, a sequential release of planning sites was required. A rule of precedence followed which on the one hand would answer the current demands and on the other hand would leave open a long-term planning option.

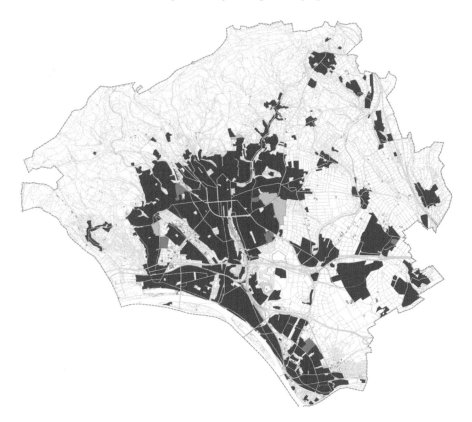

Fig. 14 Efficient land-use in urban development [28: 450]

The realization of the concept of structure of housing is geared to the following central components:

- Space (location and size of areas for additional housings)
- Time (development of areas within the scope of an aspired period of planning and a rule of precedence of development of areas)

Beyond this an oversupply of areas to accommodate demand was to be controlled and preferably avoided.
Therefore the following aims were targeted:

- Preferential use of especially suitable areas.
- Submitting an offer of a variety of areas for housing and their surroundings while also considering suitability in general.
- Supporting city centers and sub centers according to the basic objective of "Polycentrality" which correspond to Wiesbaden's historical development of housing and also concentration of the development in central city according to the basic objective of the "Compact City".

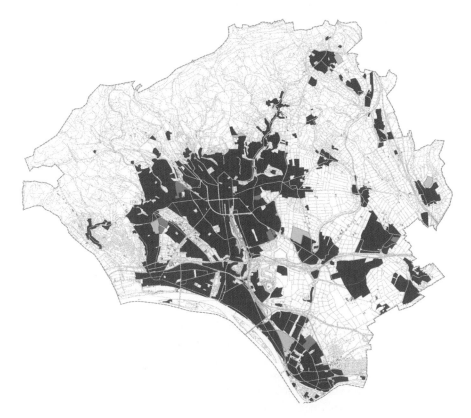

Fig. 15 Polycentric urban development [28: 452]

- Giving leeway to major projects of town planning with positive synergies as well as to minor projects.

There are two alternatives as a result of the concept of structure of settlement that connect the basic objective of the "Compact City" with the "Polycentral City".

4 Best-Practice: Sustainable Urban Living Phoenix

4.2 Framework

The second example with the title "Sustainable Urban Living" tries to transcribe the criteria of a sustainable urban development in Germany to a fictitious problem in the United States, referring to a precise spatial dimension.

4.1.1 Special Features in Urban Development and State of the Sustainability Discussion in the US

In the United States, urban development performs under circumstances that differ from those in Europe. High pressures on urban development in connection with an availability of ground and low costs for energy help to improve new housing districts in the peri-urban regions around the cities. This leads to an uncontrolled development with numerous negative social, ecological, and economical concomitants, relating to "Urban Sprawl". Following this development and facing highly decentralized patterns of land use, city centers in the United States have to cope with an imminent loss of functions and residents moving to the suburban areas accompanied by a social downgrading of districts.

Despite the intensified discussion about the effects of the urban sprawl on cities and peri-urban areas, this topic doesn't seem to have any impact on politics or even the population. The "American Dream" – representing individual freedom and the opportunity of personal development – is the major driving force [14: 21].

President Clinton founded the "Presidents Council on Sustainable Development" (PCSD) in 1993, which had to transcribe the thoughts of sustainability on the US. These were presented in 1996 in the report "Sustainable America: A New Consensus", a guideline with national aims for sustainability. However, the discussion about sustainability didn't initiate any practical changes in urban planning up to now.

An advocate of the reorientation in settlement and land use is the "New Urbanism – Movement", which tries to countervail urban sprawl with different concepts. Urban density, the composition of land use, surroundings that support pedestrians, and a good accessibility to public transport are factors for the concepts of this "New Urbanism", e.g. "Smart Growth", "Transit Oriented Development" (TOD) or "Traditional Neighborhood Design" (TND) [5: 101–109].

4.1.2   The Initial Situation in Phoenix

In trying to identify an adequate space for the survey, the city and region of Phoenix, AZ was chosen. The main characteristics of urban agglomerations in the United States were found here shaped perfectly.

Fig. 16 United States of America, State of Arizona and Phoenix [own design]

The city of Phoenix faces an immense increase in population. The number of inhabitants rose from 2.1 million in 1990 to 3.6 million today [39]. This development put pressure on the housing supply and land use that was already coping with an enormous consumption of land and energy. As a result of this, urban districts showed up with a low density in housing and population. The strong urban expansion with its decentralized structures pushed these problems in the city centers further.

Tempe, AZ, a city connected to Phoenix with its 165.000 inhabitants [41] is also affected by this overall development. The city is mainly embossed by the campus of Arizona State University (ASU), the second largest university in the United Sates.

4.2   Theoretical Identification of Sites Regarding the Principles of Sustainability

Starting from the main aim of sustainable urban land use development limiting the growth of cities and reducing the negative impacts on the ground an inner-city-area was detected. Consequently pre-used land showing all the criteria of a functionally

Fig. 17 "Urban sprawl" in Phoenix [own picture]

and practically under-used housing district between the city center of Tempe and the campus of ASU was detected.

A housing district was chosen to be activated, renewed, and redeveloped under the conditions of sustainability to be a hinge joint, set between the campus and the city center of Tempe to promote urban development.

Fig. 18 prototypical development site [Microsoft LiveSearch]

Fig. 19 impression of the prototypical development site [own image]

The main criteria in this case were the spatial connections to basic infrastructures and the availability of public transport for a TOD, supporting an efficient public transport network on one hand, and presenting an alternative option to individual automobile traffic.

Nevertheless this project wanted to give some dynamics to mesh academic research and local companies to use synergies in theory and practice to improve.

4.3 Transcription of Principles of a Sustainable Urban Development

Before beginning the redevelopment, so called "Stadtbausteine" – single components representing the tasks for urban development – were defined, having a direct impact on the whole *conceptional process*. This process refers to the complexity of tasks and detailed levels of parameters influencing urban development.

The following step was to adapt each of these "Stadtbausteine" like "Urban Area", "Transport/Infrastructure", "Urban Identity", "Public Space", "Reflection of Demographic Changes", "Mixed-Use Development", "Social Inclusiveness" and "Energy-Efficiency" to basics of the *conceptional principles* which are dedicated to the ecologic, economic, and social principles of sustainability and correlate in an alternating connection.

The main characterizations of the "Stadtbausteine" are described briefly here.

4.3.1 Urban Area

In order to facilitate a sustainable urban development it is necessary to consequently connect the project into its urban surroundings. What demands, potentials,
and restrictions follow this planning context? Which functions for the city does the
site have today and which ones should it have in the future? To us, the answers on
these and even further questions in this context seem to be quite essential for the
redevelopment and restructuring of pre-used inner city areas like in this example,
to create sustainable urban structures successfully. Leading conceptional factors are
the existing technical und social infrastructures and the structures of buildings and
functions in the (im)mediate vicinity.

4.3.2 Transport/Infrastructure

The infrastructure of transportation and supply is the skeletal structure of the city
and the city quarters, including the supply of fresh water, energy, gas and telecom-
munication services, the disposal of sewage and waste and also the access to public
transport. The aims in infrastructure in this example are maximized equality of all
traffic modes and minimized short routes of transportation within the city itself.

A fundamental question in this case was if car traffic could be reduced within the
examined area or if even freedom from cars could be an option [33]. Efficient access
to public transport, closeness to social, cultural and daily services to the ASU campus
could help reduce car traffic, hence decrease noise exposure, air pollution and CO_2
emissions, and also enhance the quality and security of public spaces.

4.3.3 Urban Identity

Urban identity correlates with the uniqueness of space and is one component in a
sustainable urban development showing a growing importance for the acceptance of
urban structures today and in the future. Regarding the general conditions in Tempe
and the intended excess planning of this district we would propose to offer urban
qualities and identification opportunities to the future inhabitants. A positive image
of this urban restructuring is on the one hand an attractive site in the inner city, while
on the other hand keeps it as an alternative to the "urban sprawl". Besides, factors
like building structure, urban density, dispersion of utilization patterns or design and
acceptance of public space, social factors like cultural, social or ethic heterogeneity,
or the offer to participate and design individual spaces play a prominent role.

4.3.4 Public Space

Streets, public spaces, and gardens emboss the face of a city dramatically and have
an enormous influence on its own identity and the quality of life. In Germany

public spaces are used for communication and interaction. Political and social life is localized here and they are integrated elements in well-working stable social urban structures.

Keeping this in mind, we proposed to declare public spaces in the US as one key-issue in the focus of a sustainable development. Here, public space is often used for transit. In contrast to that, public life is concentrated in shopping malls. Transcribing this to the example in Tempe is the generation of attractive public spaces that can also be used for identifying the site as a district allowing urban life.

4.3.5   Reflection of Demographic Changes

The demographic change is gaining more importance in spatial planning. The shifting demands of an aging society find themselves in the design of urban spaces or in the changing building stock offers. Flexibility in housing, a heterogenic pattern of utilization, open, barrier-free floor plans will be essential for planning and constructing in the future. The task in this particular case was to react to the specific ageing structure in Tempe. The city region of Phoenix, AZ, has some special climate conditions, which are attractive to many elderly people from all over the US. Many residential areas build exclusively for this target group shows this development perfectly.

Having the ASU campus near nearby the new site could help break the disconnection of age classes by creating a vivid district in which the connection of different generations was possible.

4.3.6   Mixed Use Development

Essential criteria for sustainable urban developments are heterogenic patterns of land use e.g. density, flexibility of urban spaces, and a mixture of features. Shuffling work, living, daily supplies, and recreational activities give the opportunity to live an urban life. Connecting working and living side-by-side helps organize daily life and is a postulate for a "city of short ways".

The mixture of different features is not only restricted to horizontal levels but also included in the vertical dimensions within buildings.

4.3.7   Social Inclusiveness

A heterogenic social mixture also belongs to a well-balanced patterns of use. Regarding the author's approach, different social levels and age groups, and also a mixture of different cultures should live in the new district. Heterogeneous city districts are able to regenerate more than monostructures. Stable private social networks keep the tolerance and force integration possible in Tempe to develop stable neighborhoods. This stability is a return from long-stay inhabitants being the backbone of whole districts that create a contrast to a social and economic efflux, mainly supported by highly mobile social classes.

4.3.8 Energy Efficiency

In times of limited energy resources an energy-consciousness in planning and construction becomes even more important for urban development. Protecting the environment and saving the resources is financially rewarding for the house builder and owner. Energy efficiency therefore is not only connected to ecology but furthermore to an economic aspect of sustainability. To increase energy efficiency of buildings, solar power, compact urban architecture, green roofs, solid building, construction, disconnecting thermal bridges, efficient insulation, and minimized shades are used.

Especially in the housing development in the US great opportunities for saving site and energy can be found. The considerations in this example to transcribe the conceptions of a sustainable urban development to the city and region of Phoenix, AZ, are the utilization of its various – e.g. climatic – opportunities: Phoenix, in the northern part of the Sonoran Desert, one of the largest deserts in the world, counts 330 sunny days a year [39].

These circumstances put quite a lot of strain on the settlement and housing development in the city region of Phoenix, AZ, which is tightly connected to technical solutions in building climatization. By developing new urban structures, new typologies of buildings, or by increasing the efficiency of solar power modules, the opportunities for an energy-efficient urban development can expand.

4.4 Implementation of the Defined Principles

4.4.1 Keynote to the Concepts

The defined *conceptional principles* will be applied prototypically to the site in Tempe, AZ, working on two different layouts:

The first layout tries to combine the requirements of a sustainable development with criteria of American housing to find out how and which aspect of sustainability can be imported into this typical urban development.

In the second layout the site is developed with principles of European concepts in planning, showing a possibility to plan sustainable under the local circumstances, by re-using and upgrading inner city districts.

Both layouts are based on a sensible use of climatic resources having high temperatures during the year and low rainfall. The buildings follow the given environmental restrictions. Houses with a courtyard or an atrium as well as townhouses seem to be appropriate because they have an outward sunscreen and send shadows

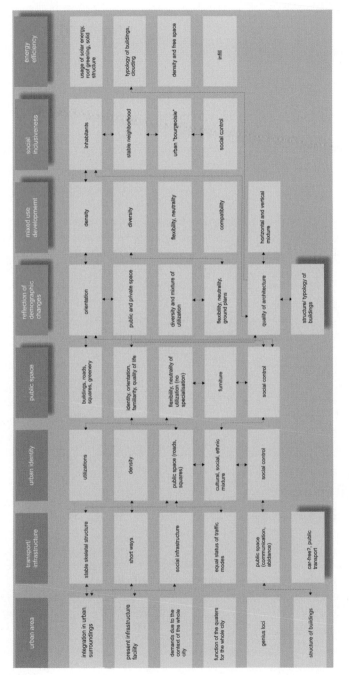

Fig. 20 "Stadtbausteine" and conceptional principles [own design]

inward. These measures are accompanied by the installation of sunscreen-glazing, solar power system, and re-using sewage.

4.4.2 Sustainable Urban Living, Concept 01

As mentioned concept 01 is orientated to typical American housing districts. The monostructured site shall be repealed by the flexibility of town houses. By mixing different urban densities livened up with public spaces many new possibilities for structuring are given. The single buildings are variable in use and assembly so that they can be combined or converted for living or working. This principle supports the optimization for different user groups and the modular floor plan

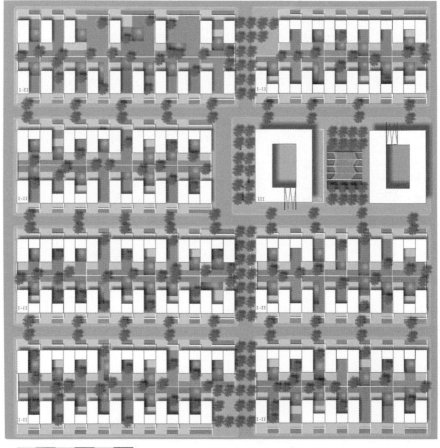

0   10   20   30   40   50 METER

Fig. 21 Draft proposal-concept 01 [own design]

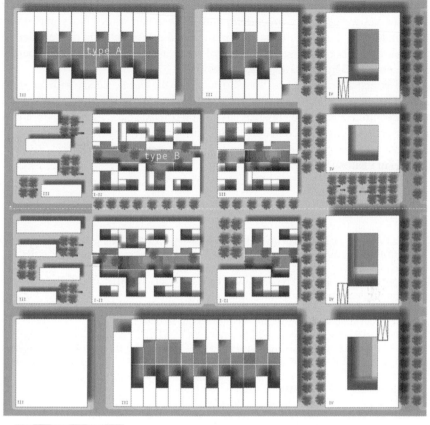

0   10   20   30   40   50 METER

Fig. 22 Draft proposal – concept 02 [own design]

gives a maximum freedom in adaptability to future demands for living and working.

Traffic-calming devices in residential roads, public spaces, and the projected shopping mall as the center underline the urban character of the district and connect it to the surroundings.

4.4.3   Phoenix – Sustainable Urban Concept 02

This concept is adjusted to the European urban development principles with heterogenic land use patterns and different typologies of buildings. The aim is to create a modern inner city district with a broad variety of uses and an adequate mixture of social groups.

The site is structured in a protected, possibly car-free residential area with courtyard houses next to a greenfield, and another part with blocks and several opportunities for shopping on the ground floor and apartments and office spaces on the upper floors. For keeping the residential area car-free, the blocks in the southern part are enhanced by a parking site.

As a connecting element to the surrounding, the public space with its traffic reduced roads and squares contribute to the high quality in living, work and life.

5   References

5.1   Literature

[1] Akademie für Raumforschung und Landesplanung (ARL). Raumwirksame Aspekte der Virtualisierung von Lebenswelten. Positionspapier zum projektierten Arbeitskreis der ARL. Hannover. 2007.

[2] Akademie für Raumforschung und Landesplanung (ARL). Handwörterbuch der Raumordnung. Hannover. 2005.

[3] G. Albers. Stadtplanung. Wissenschaftliche Buchgesellschaft. Darmstadt. 1988.

[4] AST Plan GmbH and Prof. Dr.-Ing. G. Steinebach. Siedlungsentwicklung Wiesbaden 2020. Selbst verlegt. Kaiserslautern. 2005.

[5] H. Bodenschatz and B. Schönig. Smart Growth – New Urbanism – Liveable Communities. Programm und Praxis der Anti-Sprawl-Bewegung in den USA. Verlag Müller + Busmann. Wuppertal. 2004.

[6] Bundesamt für Bauwesen und Raumordnung (BBR) (Hrsg.). Die deutsche EU-Ratspräsidentschaft – Impulse für die Stadt- und Raumentwicklungspolitik in Europa. Heft 7/8. Bonn. 2007.

[7] Bundesregierung. Nachhaltige Stadtentwicklungspolitik – ein Gemeinschaftswerk. Städtebaulicher Bericht der Bundesregierung 2004.

[8] Bundesregierung. Perspektiven für Deutschland. 2002.

[9] D. Cordes and W. Seebacher. Ostertor. Bremerhaven. 1987.

[10] R. Cromarty. Environmental Planning. Sustainable Planning. Arizona State University. Tempe. 2006.

[11] Deutscher Bundestag. Perspektiven für Deutschland. Unsere Strategie für eine nachhaltige Entwicklung. Drucksache. Kettler Verlag. 2002.

[12] M. Eltges and M. Nickel. Europäische Stadtpolitik: Von Brüssel über Lille nach Leipzig und zurück. In: Bundesamt für Bauwesen und Raumordnung (BBR) (Hrsg.) (2007). Die deutsche EU-Ratspräsidentschaft – Impulse für die Stadt- und Raumentwicklungspolitik in Europa. Heft 7/8, S. 479–486. Bonn. 2007.

[13] R. Ganser. Quantifizierte Ziele flächensparender Siedlungsentwicklung im englischen Planungssystem – Ein Modell für Raumordnung und Bauleitplanung in Deutschland?. Dissertation am Lehrstuhl Stadtplanung, Prof. Dr.- Ing. G. Steinebach. TU Kaiserslautern. 2005.

[14] B. Hagen. Anforderungen einer nachhaltigen Siedlungsentwicklung in der Metropolregion Phoenix unter Berücksichtigung des ÖPNV. Diplomarbeit an der TU Kaiserslautern, Lehrstuhl Stadtplanung. Selbst verlegt. Kaiserslautern. 2007.

[15] Hessisches Ministerium für Wirtschaft, Verkehr und Landesentwicklung. Demographische Rahmendaten zur Landesentwicklung. Selbst verlegt. Wiesbaden. 2004.

[16] C. Jacoby. Die Strategische Umweltprüfung (SUP) in der Raumplanung. Instrumente, Methoden und Rechtsgrundlagen für die Bewertung von Standortalternativen in der Stadt- und Regionalplanung. 2000.
[17] H. Kistenmacher et al. Ermittlung des Wohnbaulandpotenzials in Verdichtungsräumen unter besonderer Berücksichtigung der Umweltverträglichkeit. Untersucht und dargestellt am Beispiel des Verdichtungsraums Stuttgart. 1988.
[18] R. Kreibich. Städte in Deutschland: Herausforderungen für die Stadtentwicklung. In: Bundesministerium für Verkehr, Bau und Stadtentwicklung (2007). Auf dem Weg zu einer nationalen Stadtentwicklungspolitik – Memorandum. S. 53–55. Berlin, Bonn. 2007.
[19] Presidents Council on Sustainable Development. Sustainable America. A new Consensus. Washington. 1996.
[20] D. Reinborn. Städtebau im 19. und 20. Jahrhundert. Stuttgart, Berlin, Köln. 1996.
[21] K. Selle. Öffentliche Räume in der europäischen Stadt – Verfall und Ende oder Wandel und Belebung. Reden und Gegenreden. In: W. Siebel(Hrsg.). Die europäische Stadt. Edition Suhrkamp 2323. Suhrkamp Verlag. S. 131–145. Frankfurt am Main. 2004.
[22] W. Siebel. Was ist die europäische Stadt?. In: EuropaKulturStadt. Deutscher Kulturrat e. V. Ausgabe II, Januar/ Februar. S. 1–2, Frankfurt. 2005.
[23] W. Siebel Die europäische Stadt. Edition Suhrkamp 2323. Suhrkamp Verlag. Frankfurt am Main. 2004.
[24] Stadt Kaiserslautern. Pilotvorhaben Lärmminderungsstudie Innenstadt-West. Prof. Dr.- Ing. G. Steinebach, Lehrstuhl Stadtplanung, TU Kaiserslautern. 2005.
[25] Stadtplanungsamt Wiesbaden. Siedlungsentwicklung Wiesbaden 2020. Selbst verlegt. Wiesbaden. 2004.
[26] Statistisches Bundesamt (Hrsg.). Bevölkerung Deutschlands bis 2050 – 11. koordinierte Bevölkerungsvorausberechnung. Wiesbaden. 2006.
[27] G. Steinebach. Raumrelevanz der Virtualisierung. In: Europäische Akademie zur Erforschung von Folgen wissenschaftlich-technischer Entwicklungen. Bad Neuenahr-Ahrweiler GmbH (Hrsg) (2007). Gesellschaftliche Randbedingungen der Virtualisierung urbaner Lebenswelten. Graue Reihe, Nr. 42. S. 9–28. 2007.
[28] G. Steinebach. Siedlungsentwicklung Wiesbaden 2020. AST GmbH and Prof. Dr.-Ing. G. Steinebach. Kaiserslautern. 2005.
[29] G. Steinebach. Im Herzen das Mittelalter, im Kopf die Stadt des 21. Jahrhunderts. In: Die Rheinpfalz. Nr. 132, 09.06.2004. 2004.
[30] Umweltbundesamt (Hrsg.). Lokale Agenda 21 im Kontext der kommunalen Steuerungsinstrumente auf kommunaler Ebene. Berlin. 2002.
[31] United Nations Organization (UNO). Brundtland Report. Our common future. Report A/ 42/ 427. 1987.
[32] The Cities Alliance. Liveable Cities – The Benefits of Urban Environmental Planning. Washington D.C. 2007.
[33] Arizona State University, School of Planning. Phoenix Camelback Corridor: Transit Oriented Development, Urban Planning Studio. Phoenix. 2007.

5.2   Internet

[34] @BBR.http://www.bbr.bund.de/cln_005/nn_115406/DE/ForschenBeraten/Stadtentwicklung/StadtentwicklungDeutschland/Tendenzen/Rueckblick/07__Stadterneuerung_20in_20den_20 70er_20Jahren.html. Abruf: 04.12.2007.
[35] @Bochum. http://www.bochum1.de/Bilder/Leipzig2004/Einkaufsarkaden_Leipzig_ Hauptbahnhof_gross.jpg. Abruf: 29.05.07.
[36] @DESTATIS. http://www.destatis.de/download/d/ugr/suv2005.pdf Abruf: 11.04.2007.
[37] @Leipzig. http://www.leipzig-online.de/index.php3/81_1_1.html?g=11. Abruf: 04.12.2007.

[38] @Nachhaltigkeitsrat. http://www.nachhaltigkeit.info/artikel/geschichte_10/rio_48/weltgipfel_
      rio_de_janeiro_1992_539.htm. Abruf: 04.12.2007.
[39] @Phoenix. http://phoenix.about.com/cs/living/a/PhxFastfacts01.htm. Abruf: 05.12.2007.
[40] @Planungsverband. http://www.planungsverband.de. Abruf: 04.12.2007.
[41] @Tempe. http://www.tempe.gov. Abruf: 05.12.2007.
[42] @Wiesbaden. http://www.wiesbaden.de. Abruf: 04.12.2007.
[43] @Wordsummit 2002. http://www.worldsummit2002.de. Abruf: 04.12.2007.

Visualizing Planning for a Sustainable World

Subhrajit Guhathakurta

Contemplating Sustainable Planning

It is easy to recognize unsustainable development, and such forms of development are all around us. The fast diminishing tropical and temperate forests in the Amazon and Southeast Asia, the growing appetite for fossil fuels leading to both depletion of petroleum reserves and increasing global warming, the hazy air filled with particulate pollution in Beijing and Mexico City, and the burgeoning waste streams making its way to our fresh water repositories on the surface and below ground all remind us about a socio-political-economic system that is heading towards disaster. Sustainability has become a guiding principle to avert this bleak future. It is also a commitment to transforming our lifeworld in a manner such that the future generations can also enjoy the bounties of nature and maintain a high quality of life.

The current context of sustainable development has entered a new phase that is fundamentally altering the earlier debates about the concepts and definitions of sustainability. Sustainability emerged in the late 1970s as an ethical concern for the ability of future generations to at least maintain the quality of life enjoyed by the present generation. Although the principles of sustainability are well known, planning for sustainable development has been the subject of considerable debate. Given vast existing inequities in wealth, power, income, and quality of life, privileging the needs of future generations over the current is problematic. Therefore sustainability has become a politically charged word that encompasses all core issues of development. As Anand and Sen (2000) very poignantly observe "… this goal of sustainability, increasingly recognized to be legitimate, would make little

S. Guhathakurta
School of Planning and Global Institute of Sustainability, Arizona State University, Tempe, AZ 85287-2005, USA, E-mail: subhro@asu.edu

G. Steinebach et al. (eds.), *Visualizing Sustainable Planning*,
© Springer-Verlag Berlin Heidelberg 2009

sense if the present life opportunities that are to be 'sustained' in the future were miserable and indigent. Sustaining deprivation cannot be our goal, nor should we deny the less privileged today the attention that we bestow on generations in the future." (p. 2030).

Ultimately the quest for sustainability has remained firmly within the confines of our current paradigm of privileging economic growth over all else. The principles of sustainability do not exhort us to make fundamental changes in our lifestyle choices, rather, we are asked to be "smarter" about doing what we already do. That is, being more efficient about the use of scarce resources – perhaps by reducing unnecessary energy use and by using water and energy conserving appliances and fixtures. We expect marginal changes to our lifestyle to make a significant difference to sustainable outcomes.

Global climate change (GCC) has shifted this debate from the issue of ethics and local degradation to the possibility of worldwide catastrophe. It is now more about the survival of the planet and many of its inhabitants than about localized issues of toxic waste streams and resource scarcity. The debate transcends the ethical dilemmas about where we should place our concerns while bringing into sharper focus other more urgent problems that are global in nature. This has led to the recognition that profligate economic growth cannot be sustained especially if most inhabitants tend to consume similar basket of goods as those in developed nations. There is also increasing evidence about the likelihood of most impacted populations to be in regions that have contributed least to greenhouse gas emissions and have the least ability to adapt. Hence, we can no longer assume that marginal changes will put us firmly on the road to sustainability, especially if everyone else wants to achieve the same economic status as those of us in the developing world. And more significantly, we may not have time for a technological bailout of this impending cataclysm.

Pedagogic Urban Models for Sustainable Planning

Given the impending challenges posed by unsustainable growth and the specter of climate change, how should we plan our cities and communities? Are the current planning techniques and processes adequate to address the issues discussed above? As planners we are expected to confront these questions and suggest an appropriate way forward. Also, in forging the strategies for planning with climate change we need to keep in mind the core values in planning that include a focus on equity, efficiency, a long-term horizon, and community participation.

Planning for a sustainable future inevitably requires some knowledge of the range of issues that could define that future. Developing a knowledge base for possible GCC related impacts is an essential starting point for any planning in this era of climate change. The knowledge base also provides some hypothesis and benchmarks from where a large number of scenarios can be generated. These scenarios are not just "projections" but informed speculations that emerges from an exercise in simulation modeling and participatory brainstorming. The process is self-reflexive and critical – a requirement imposed by the fact that the existing empirical tools could fail to uncover significant events. What is needed is a new class of models that serve as learning tools and allows us to explore the potential impacts in space and time.

While operational urban land use and transportation models have made some-what of resurgence in the past decade, the structure and function of these models have not strayed too far from the deterministic flavor of the first-generation models. This chapter is an attempt to travel along the path from deterministic models to models of dynamic systems and self-organization. It evaluates the functional aspects of the various approaches to modeling and shows that by moving our frame of refer-ence from prediction to exploration and speculation we are able to transition from future state projections to speculative theory building. Given that our knowledge of socio-behavioral impacts on urban form is still limited, the use of models for specu-lating about sustainable development is a more critical enterprise than the applica-tion of models in various contexts. Urban models within this frame of reference become a tool for clarifying mental models, for witnessing unexpected emergent properties, and for advancing planning communication, pedagogy, and ultimately, sustainability.

The paradigmatic transition of modeling approaches is especially important given that socio-behavioral systems are composed of individual elements capable of learning, adapting, and transforming. Hence, the assumption of a time invariant and unchanging underlying condition that drives some current urban models may be problematic, especially in long-range predictions. Sustainable urban develop-ment in this context becomes the organic evolution of a series of events that are sparked by small and limited interventions in highly sensitive processes. Choosing where to intervene becomes a critical exercise, which can be accomplished with the help of dynamic simulation models. Effective planning would address the most sensitive and critical drivers of sustainable human behavior as determined through such models.

The objective of this exposition on pedagogic aspects of models is not to dis-credit any form of modeling but to draw bridges between them. The focus is on the evolutionary aspects of concepts, theories, models, planning practices and peda-gogy. It is also an acknowledgment of the proposition that the development of ideas and approaches are the part and parcel of the epoch in which they formed and matured. Hence, all conceptual models address some need to resolve conceptual or practical problems and is able to clarify these problems if not resolve them. If in fact these models have served in framing problems, clarifying interrelationships, and exposing indeterminacies, they have certainly been useful. It is through such critical examination of models that we discover many conceptual inadequacies and incomplete or even erroneous descriptions of phenomena. Therefore, the use of models for exploring as well as communicating (theoretical or mental) constructs is an essential evolutionary process. All modeling endeavors should be perceived as part of this evolutionary process.

A second objective of this chapter is to draw attention to the use of models in policy-making. It is in the context of policy-making that the pedagogic value of models assumes an important role. The term "pedagogic" is used in the broader sense of concepts and tools that are "knowledge imparting". This is in contrast to the more common use of urban models in futures assessment and scenario development, which focuses more on the resulting future state the model predicts than on the adequacy of our knowledge in comprehending the extant reality. In their pedagogic role, urban models illustrate the essential evolutionary logic and theoretical constructs that tie together complex elements of that extant reality. Hence pedagogic models reconstruct and communicate our mental constructs (formed by both formal and informal learning processes) about relationships and evolutionary pathways of objects consisting of a) decision-making agents; and b) stationary and non-stationary objects in the space-time environment; within a bounded space-time framework.

A central concept in the pedagogic view of urban models is the character of space and time, or rather space-time. As in the classical distinction between the mechanical and the quantum laws of physics, the conceptual variation in space-time in social processes do not necessarily confront many contradictions as long as we stay within the limited framework assumed by our mental constructs. Newton's laws of motion and gravitation are adequate in determining with a high level of accuracy the trajectory of a baseball struck with a given force at a given angle. However, the same laws when applied to a moving object in distant space would fail to provide accurate results. Similarly, social processes can and have been modeled at different spatial and temporal scales with varying levels of success. As long as our conceptual limitations are acknowledged and exposed, the usefulness of models can be assessed for some aspects of the modeling process.

Reconceptualizing Space-time in Urban Models

The concept of space and time has been separated in our consciousness as a result of the mechanistic view of the universe in the intellectual tradition of Kant's "Transcendental Aesthetic" (Kant, translated by Smith 1965). While Kant assigns to space an empirical reality outside of the subjective (an a priori condition for intuitively assigning and delineating objects), time is considered to be a purely subjective condition of human intuition. Time, apart from the subject, has no existence in the transcendental context. Regardless, Kant acknowledges that in respect to all appearances, and therefore of all the things, which can enter into our experience, time is necessarily objective. The critical issue for the empiricists has been the separation of space from its contents and putting such objects back in space intuitively, a feat that is only accomplished by assuming the objective reality of time.

Time was wished away in the nineteenth century through the search and conceptual acceptance of immutable general laws. The ideal of science was in reaching an unchanging eternal reality in which the past, present, and the future would have exactly the same influence on the laws of nature. IlyaProgogine describes this ethos thus: "One of the main fascinations of modern science was precisely the feeling that science had exorcised time, that it could be formulated entirely in terms of basic eternal laws in which no reference to time need ever be made" (1985, p. 5). In fact the unchanging, homogeneous aspect of time imparted a feeling of intellectual security given that it disallowed the existence of the eternally unknowable (Levy-Bruhl quoted in Prigogine 1985). Since in this context the past and the future impart identical influence on natural laws, they are interchangeable. Accordingly, intellectual security rested on the fact that the unknown is only an artifact of our current ignorance and would be resolved once, at some point in the future, we discover the underlying laws that would apply uniformly to the homogeneous continuum of time. It was only with the advent of relativity theory that science began to confront the problematic posed by the multiple dimensions of time.

The mechanistic view of time has left its imprint on the social sciences in a profound way. In the models that made their way from the natural sciences to the social sciences, the temporal dimension is clearly absent as in the "spatial interaction" type models. Similarly, the dominance of econometric concepts of static and general equilibrium implicitly considers time to be either a unitary moment (as in the point of equilibrium) or a homogeneous, one-dimensional entity (as in the short and long run demand/supply charts). Further, there appears to be a temporal lag for ideas and concepts, which cross over from the natural to the social sciences and continue to persist well beyond the point when paradigms in the natural sciences have shifted. Despite the temporal lag, the conceptual framework for comprehending human-environment relationships has evolved over time and new approaches have incorporated dynamics in both space and time. Interpreting and communicating the epistemological constructs in these approaches, however, need to be carefully distilled and distinguished from the previous atemporal paradigm. An essential aspect of this act of distinguishing the paradigms requires us to understand our consciousness of and perception of time.

The problematic posed by time as a subjective condition troubled Einstein who admitted that there is something essential about "Now", which is outside the realm of science. For Einstein time acquires an objective meaning only when it can be measured and communicated from one observer to another (Prigogine 1985). It is curious that his work led to the discovery that under some circumstances, the communication of time between observers occupying different situations is impossible. Relativity theory denies the possibility of a "universal now", therefore, there is no borderline separating the past and the future. It is suffice to say in this exposition without extolling the intricacies of relativity theory that in the relativistic universe there are sequential states of experience "given as one block" of four-dimensional space-time (trajectories encompassing the totality of states assumed by objects during their lifetime). The temporal dimension is outside this block and supervenes on it through our mental (subjective) construct. (Franck 2003) provides an excellent metaphor for understanding the complex and often confusing dimensions of temporal change. He constructs an image of an "ocean of world states" in which the "now" is a wave that rolls through this ocean at a certain speed. This image accounts for several features characteristic of our experience of time: a) the image suggests that the "now" persists while the world states (including the inner states of consciousness) come and go; b) the image also evokes the perception that "now" is in perpetual and relative motion; c) hence the image highlights the ambiguity of time where "now" is simultaneously persistent and in relative motion. Given such ambiguities, empirical methods that incorporate time require some critical assumptions, which are usually glossed over in current urban models.

As conscious beings, we are aware of our presence in time and hold images of the past as well as the future. Therefore, temporal change does present certain components that are amenable to empirical methods[1]. If the concept of time is a mental process then the most appropriate approach for codifying time empirically would be a first person account of what happens when time goes by.

It is only in the human mind that the relationship between time, as measured by the distance moved by the "now wave", and time as experienced is established. In other words, there is a distinction between the measure of time as distance (which can be metered by clocks) and the duration, which took for the same time to pass as experienced. This distinction is succinctly explained by (Franck 2003) thus: "Clocks do not measure the passage of time but translate distances in time into distances in space. Duration, in contrast, does refer to the now. Our sense of duration is the sense that the now endures while the time slices being raised to presence come and go" (page 95).

The two states of time and duration can be described in a chart presented in Fig. 1 in which time (t_i) is related to a discrete quantum of "nowness" (η_i). As t_i moves along the axis t, it simulates the shifting of "now" as in the wave in the "ocean of world states". The "now" can also be viewed as the integration of the sequence of quanta $\Delta\eta_i$, which gives the impression of a persistent "now". In essence, t_i measures clock time (as distance) while η_i measures duration. Although the two measures are related, they are not identical. More importantly, the perception of duration is intersubjective and can only be described in the first person (Franck 2003).

The perception of time in the context of sustainability significantly highlights several aspects of subjectivity noted by Franck (2003). First, the concept of

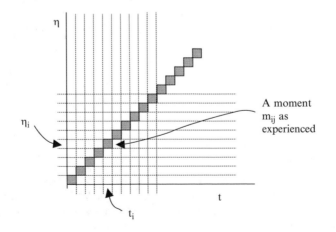

Fig. 1 Sequence of moments defined by clock time and "nowness"

sustainability attempts to overcome time and its associated uncertainties by focusing attention on "now", that is, on our current practices and modes of behavior. Second, the idea of "steady-state" espoused by some ecological economists like Herman Daly (1977, 1989) mirrors the mechanistic view of the universe described earlier by emphasizing basic laws that are considered timeless. Third, the idea of rights of the unborn, and our obligations to future generations enshrined in most principles of sustainability is uniquely a construct of our present attitudes and values that are projected to encompass those who are not present. Although the notion of sustainability embraces many generations in the future but it subsumes the ethos of the present. In many respects, we are reliving the debates about the search for atemporal basic laws that might lead us to a "steady state" on one hand and on the other an acknowledgement of the "arrow of time" and the path dependency of the actions we take today. The dynamic spatial modeling approach described below can illuminate the assumptions behind both sides of the debate about the concept of time in thinking about sustainability.

The Dynamic Spatial Approach to Land Use Modeling

In a truly dynamic land use model, land use changes unfold as a sequence of events determined by the interaction of forces incorporated in the model at predetermined intervals. At each interval, the model checks for parameters that may have changed during the last interval and recalculates the impact on all objects included in the model. These objects consist of decision-making agents as well as other environmental factors that restrict or enable change of spatio-temporal states. In contrast to the empirical-statistical approach described above, the relationships in a dynamic model can be multi-directional. One common form of dynamic simulation approach is found in the system dynamics models popularized by Jay Forrester (1969, 1971), Meadows et al. (1972, 1992), among others. However, there is more to dynamic simulation than system dynamics. Dynamic simulation can include a wide range of models such as Monte Carlo (stochastic) methods, object based programming, and cellular automata models. The dynamic land use model described below incorporates both system dynamics and cellular automata based methods.

I provide an example of modeling land use change for a place in southern Arizona to highlight the evolutionary and the pedagogic aspects of simulation models. The place is Sierra Vista in Cochise county, Arizona, which has been studied extensively in the literature (e.g., Guhathakurta 2002, Guhathakurta 2004; Steiner et al. 2000). The schematic of the dynamic land use change model for Sierra Vista is provided in Fig. 2. This model has two essential components: 1) the aggregate growth in demand for land; and 2) the allocation of growth to discrete parcels most suited to residential development. The interaction between growth and demand for land is mediated by the sub-model called "demand-supply match" that matches demand for and supply of residential land. As will be discussed shortly, demand and supply are not expected to be in equilibrium at any

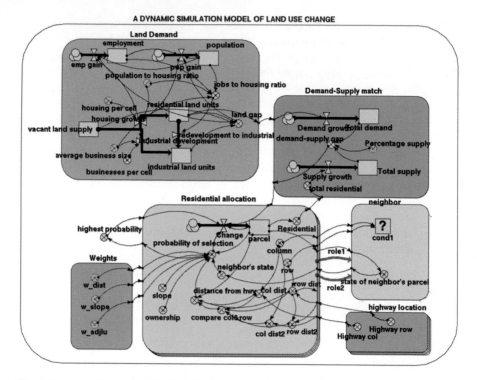

Fig. 2 Dynamic Simulation Model in Simile

stage in this dynamic model, which is in contrast to the equilibrium assumption driving most econometric models.

The dynamic land use change model is constructed in "Simile", software for designing and running complex dynamic models developed and marketed by Simulistics Ltd. Among the various advantages of Simile as software for dynamic modeling perhaps the most significant is the ability to develop and visualize complex spatial models. Simile offers the flexibility of writing complex equations and program statements for both quantitative and qualitative (categorical) data. The residential allocation component of the model shown in Fig. 2 assumes a region defined by a 20×20 rectangular grid where each grid encompasses about 20 acres. A region on the outskirts of the city of Sierra Vista was chosen because such an area would be most impacted by growth pressures. This area is also intersected by an Interstate freeway and has a mix of residential use, commercial use, and protected open space. The intention was to incorporate the most salient factors impacting the selection of land for residential growth including accessibility to highways, character of adjacent land uses, and the physical characteristics of the land (represented by the slope parameter). The sustainability indicators in this case are derived from future development patterns that are more proximate to commercial areas (mixed use) and highways as well as the protection of foothills (slope).

The residential allocation model calculates a "probability of selection" for each grid cell based on a linear combination of four factors: 1) adjacent land use; 2) distance from freeways; 3) slope; and 4) ownership characteristics. At each time-step the model calculates this probability for each cell and holds in memory the highest probability of selection of all 400 cells used in this analysis. The probability of selection for each cell is then matched with the highest probability of selection for all cells. The matched cells that are vacant and zoned residential are converted to residential at each step provided there is a demand for residential growth at that time.

The results are perhaps best viewed as a movie showing the sequence of states of residential land development. As this is not possible in this publication, four snapshots of residential states are provided in sequence in Fig. 3. What these

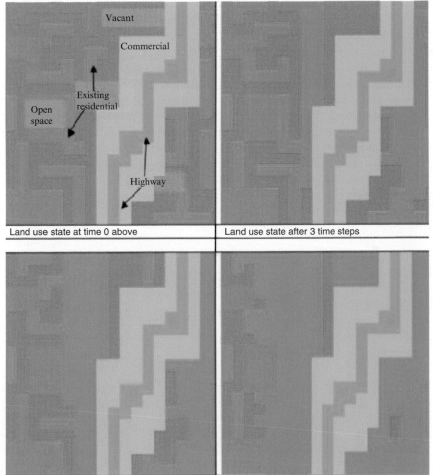

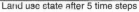 

Fig. 3 Visualizing the changing state of residential land use in time steps (years in this case)

snapshots fail to show is the fact that residential development proceeds in quantum jumps at some periods followed by several periods of lull. The figure is however able to show the clustering effect of residential development not too far from the freeways. The location is also strongly influenced by the weight assigned to the slope factor (i.e., steeper the slope, the more expensive to develop). This was confirmed by reducing the weight on slope and noticing higher levels of growth in the north-eastern part of this region that may be considered to be unsustainable both by virtue of its distance from existing services and destruction of more fragile ecosystems at the foothills.

A dynamic model such as the one described here allows for a number of theoretical exercises. For example, the model can test the proposition that the type of adjacent land use does not influence decisions of developers who develop large housing subdivisions that are often self-sufficient. Expectedly, "shutting down" the influence of adjacent land uses does not result in a random non-contiguous development. In fact the change in the sequence of residential development is almost imperceptible. This is not, however, counterintuitive given that other factors impacting residential land use decisions such as slope and accessibility changes gradually over space. In contrast, experimenting with the impact of accessibility and slope offers more dramatic changes in the sequence of residential development. Therefore this model offers an excellent way to assess the interplay of forces that impact sustainable development of residential land both spatially and temporally.

Another important feature of this spatial-dynamic model of residential development is its ability to discover unexpected emergent properties of land markets. One such property of land markets is the well-known and historically documented process of boom and bust cycles similar to the business cycles in the larger economy. Figure 4 shows the simulated cycles of excess demand followed by excess supply

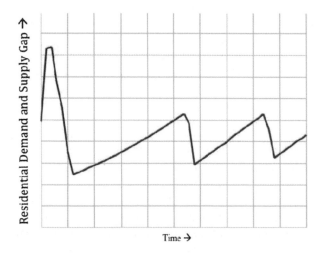

Fig. 4 The real estate development cycle as an emergent property of the dynamic spatial model

in the plot of "demand-supply gap" with respect to time. This figure confirms our observations about the real estate market although it was discovered serendipitously after the model was developed. We can now rationalize how this cycle was actually a result of discrete processes that were modeled – that is, the probability for selection accepted discrete rather than continuous values resulting in several contiguous cells receiving identical probabilities at each time-step. This is closer to reality as the decision to develop is not impacted by minor variations in distances or physical attributes. Hence the use of discrete ordinal values offers a more attractive alternative in urban land use change models.

The Pedagogy of Urban and Environmental Models

The model discussed above have some distinct qualities that offer a different perspective on urban land use change. It is important to emphasize that dynamic spatial models of the type discussed in this chapter do not have the ability to discover empirical regularities that have not been assumed a priori. For example, the relative influence of physical properties of land compared to accessibility or adjacent use cannot be "discovered" from the model parameters. Even if the more fundamental characteristics of behavior of decision-making agents were modeled and the factors impacting residential land derived from such behavior, these fundamental characteristics cannot be known without systematic empirical observation. Hence exogenous parameters, empirically derived, are essential inputs in spatial dynamic models. In contrast, stochastic-empirical models provide a robust method for determining the relative impacts of different factors impacting land use. Simulation models cannot dispense with empirical associations, which provide the range of parameters serving as benchmarks for running the model. In this respect, stochastic-empirical models provide complementary information for running the dynamic-simulation models. Therefore, both forms of models are useful and have distinct and often complementary pedagogic aspects.

It is, however, important to reject the perception of objectivity and resulting superiority of stochastic-empirical models constructed from large detailed data sets. The issue of measurement precision often seems to overwhelm our assessments of the underlying theoretical premises that are being modeled. We begin to ask the less important question about the fit of the model with the available data rather than the more significant pedagogic questions about the models' fit with our theories and vice-versa. By asking the former question we disregard the pedagogic aspects of models and accept our a priori theories uncritically. Also significantly, we fail to confront or acknowledge the uncomfortable assumptions about space and time embedded in information that is fed into such models. It is theoretically incorrect to assume a singular data point as a snapshot in time given that time cannot but be in motion. It is only through the subjective mental process that static images can be realized. Such images are, in fact, intersubjective and can be communicated through first-person accounts by assuming a priori a measure of time (clock time).

In other words, empirical-stochastic models tend to privilege observations that can be accurately described over other significant observations that may be less amenable to precise measurements.

Dynamic simulation models are less sensitive to empirical information and more to the rules of behavior that guide individual actions. Time enters simulation as an integral component but only in so far as it is associated with the sequence of states of objects (agents and their environment). Although a "clock-time" is assumed in simulations, time actually enters only as duration that is termed a "time-step". The essential structure of a dynamic simulation is not dependent on a particular time-step, although conceptually there may be valid reasons to assume duration measured in the available known metrics (calendar year, hour, month etc.). However, assumption of a particular "time-step" does not change the qualitative structure of a simulation given that a change in the time-step universally leads to slowing down or speeding up of the sequence of events. Choosing a lower or higher "time-step" may also invoke better representation of small variations, or conversely, impart a smoothening effect on the states of the system being represented. Hence, dynamic simulations offer an excellent means for examining the change of world states irrespective of the assumed duration, which may be derived subjectively or consensually.

Another attractive feature of dynamic simulation as shown earlier is the ability to observe emergent properties. Many emergent properties are difficult to describe analytically with a high degree of precision. However, by repeated observations under test conditions an appreciation of the resultant effects can be gained that may either provide valuable information for decision-making or illuminate certain properties requiring further scrutiny. In such cases the computer simulation bridges the gap between "speculative inquiry", a domain of philosophy, and empirical techniques that has dominated scientific research overthe past two centuries. The computer program serves as an analogy of an explanatory theory, which is tested within the controlled "virtual" environment and modified if the results do not conform to observed facts. The testing of simulation models may continue through a structured analytical process or through an unstructured iterative (sometimes numerical) process or both. The objective is to express, refine and test the underlying explanatory theory under specific contexts, which may lead to the application of the simulation model as essentially a pedagogic undertaking.

Given that dynamic simulations and visualization offer several advantages such as a) the ability to address unstructured problems; b) the possibility of visualizing emergent properties that are often unexpected; c) the ability to treat time with variable metrics determined subjectively or consensually; and d) the ability to communicate through narratives; it is likely that most urban and environmental models would include some aspect of simulation. Shifting the emphasis from empirical/deterministic models to simulation, without rejecting the former entirely, leads to new forms of expression that may offer a different understanding of the social and ecological evolution.

References

Anand, S. and A. Sen (2000) Human development and economic sustainability. *World Development* 28(12), 2029–2049.

Daly, H. (1977, 1991 2nd ed.) *Steady State Economics.* Washington DC: Island Press.

Daly, H., J. B. Cobb, and C. W. Cobb (1989) *For the Common Good: Redirecting the Economy toward Community, the Environment, and a Sustainable Future.* Boston, MA: Beacon Press.

Forrester, J. (1969) *Urban Dynamics.* Waltham, MA: Pegasus Communications.

Forrester, J. (1971) *World Dynamics.* Cambridge, MA: Wright-Allen Press.

Franck, G. (2003) How time passes. In R. Buccheri, M. Saniga, and W. M. Stuckey eds. The Nature of Time: Geometry, Physics and Perception. Dordrecht: Kluwer Academic Publishers. Pages 91–103.

Guhathakurta, S. (2002) Modeling as storytelling: using simulation models as a narrative. Environment and Planning B: Planning and Design 29: 895–911.

Guhathakurta, S. (2004) "Telling Stories with Models: Reflecting on land use and ecological trends in the San Pedro Watershed." In David Unwin (ed.) Re-Presenting GIS. New York, NY: John Wiley & Sons.

Kant, E. (1965) The Critique of Pure Reason. Translated by Norman Kemp Smith. New York: St. Martin's Press.

Levy-Bruhl, L. (1922) La Mentalité Primitive. Paris: PUF. McFadden, D. (1974) The measurement of urban travel demand. Journal of Public Economics 3: 303–328.

MacEachren, A. M., M. Gahegan, W. Pike, I. Brewer, G. Cai, E. Lengerich, and F. Hardisty. (2004) Geovisualization for knowledge construction and decision support. IEEE Computer Graphics and Applications 24(1): 13–17.

Meadows, D. H., D. L. Meadows, J. Randers and W. W. Behrens (1972) *The Limits to Growth.* New York, NY: Universe.

Meadows, D. H., D. L. Meadows, and J. Randers (1992) *Beyond the Limits: Confronting Global Collapse, Envisioning a Sustainable Future.* London: Earthscan.

Muller, P. G. Zeng, P. Wonka, and L. V. Gool (2007) Image-based procedural modeling of facades. Proceedings of ACM SIGGRAPH 2007 / ACM Transactions on Graphics, 26(3).

Muller, P, P. Wonka, S. Haegler, A. Ulmer, and L. V. Gool (2006) Procedural modeling of buildings. Proceedings of ACM SIGGRAPH 2006/ ACM Transactions on Graphics, 25(3): 614–623.

Parish, Y. I. H., and P. Muller (2001) Procedural modeling of cities. Proceedings of ACM SIGGRAPH 2001: 301–308.

Prigogine, I. (1985) Time and human knowledge. Environment and Planning B: Planning and Design 12: 5–20.

Steiner, Frederick. R., John Blair, Laurel McSherry, Subhrajit Guhathakurta, Joaquin Marrufo, and Matthew Holm (2000) Designing an Environmentally Sensitive Area Protection System for the Upper San Pedro Drainage Basin (Mexico and USA). Landscape and Urban Planning 49(3/4): 129–148.

Taleb, N. N. (2007). *The Black Swan: The Impact of the Highly Improbable*. New York, NY: Random House.

GIS-based Applications and the English Planning System: Modeling Town Centers and Housing Potential

Daryl A. Lloyd

Abstract The monitoring of town centers has become very important to Her Majesty's Government (HMG) over the past 15 years. This paper describes how a GIS-based tool has been devised for the defining and monitoring of English town centers and its role in the English planning system. Ultimately, the objective and empirical measurement of something as conceptual as a "town center" is not an easy task, and some of the difficulties of carrying this out are discussed.

A further proposal is put forward to adapt the town centers methodology to produce a new tool for modeling housing capacity.

The Role of Government in Planning

In the United Kingdom responsibilities for the planning system are split between three separate levels of government. These responsibilities are divided along operational and strategic lines, with local planning authorities handling the planning applications and development system, and central and regional government acting with a strategic drive.

Communities and Local Government is the central government department that takes most responsibility for planning in England. They do this by setting overall aspirations and devising policy guidance notes that give indications to the delivery bodies, such as local planning authorities, how planning should be taken forward. These notes, formerly called Planning Policy Guidance (PPG), are now referred to as Planning Policy Statements (PPS) and cover 25 different topics (see Table 1 for complete list).

Some of these notes and statements are run jointly with other departments where the responsibilities are shared. The overarching aim is to drive and influence the planning process to produce development (of both a residential and commercial and industrial nature) in line with the current policies of central government.

D.A. Lloyd
Centre for Advanced Spatial Analysis, University College London, 1-19 Torrington Place, London WC1E 7HB, UK, E-mail: daryl.lloyd@dunelm.org.uk

G. Steinebach et al. (eds.), *Visualizing Sustainable Planning,*
© Springer-Verlag Berlin Heidelberg 2009

Table 1 Central government planning guidance

Planning Policy Guidance (PPG)	Planning Policy Statement (PPS)
2 Green belts	1 Delivering sustainable development
4 Industrial, commercial development and small firms	3 Housing
5 Simplified planning zones	6 Planning for town centres
8 Telecommunications	7 Sustainable development in rural areas
13 Transport	9 Biodiversity and geological conservation
14 Development on unstable land	10 Planning for sustainable waste management
15 Planning and the historic environment	11 Regional spatial strategies
16 Archaeology and planning	12 Local spatial plans
17 Planning for open space, sport, and recreation	22 Renewable energy
18 Enforcing planning control	23 Planning and pollution control
19 Outdoor advertisement control	25 Development and flood risk
20 Coastal planning; Planning and noise	
24 Planning and noise	

The Government Office Region network (with England split into nine separate areas: London, South East, South West, East of England, East Midlands, West Midlands, Yorkshire and the Humber, North East and North West) acts as a delivery hub for ten central government departments. Working with local authorities and other bodies on planning, housing, transport, and the environment is a significant part of their work, and this includes helping translate the PPGs and PPSs into something more directly useable.

Local plans (now Local Development Frameworks [LDF]) and Regional Spatial Strategies (RSS) are both documents that fall into this area. Unlike elsewhere in Europe, in England there is a system where central government outlines the broad policies, regional government (though the government offices and regional assemblies) provide interpretation and local fine-tuning, and local planning authorities actually apply this to the "real world," with a considerable amount of autonomy devolved downwards.

The RSS acts as a region-wide document that starts to apply the guidance from the central government to a more manageable area, taking care to make consideration for the local economic, social, and geographical conditions. This is important as it can reflect any critical sub-regional areas (for instance, illustrating that a cluster of towns and villages are actually feeders for a larger town or city) and therefore through it plans can be made and adapted accordingly to account for local conditions. Other factors here can include water supply, waste disposal and transport links. All of these things influence the performance of a region and can help a region have distinct characteristics in comparison to the others. Without a regional body to do this, all co-ordination would have to be carried out by the individual local authorities, requiring up to 50 bodies to work together without anyone taking a overview. Each RSS is constructed to take a long-term view, setting out the strategy for the next 15 to 20 years, and goes through a rigorous public review process (called Examination in Public, EiP) to assure that it is as complete and fair as possible.

Each local authority then develops its own LDF, outlining exactly how government policies and the RSS on spatial planning will be applied to the local community, taking into account the local conditions. This is not one single document, but rather a bundle of development documents on community involvement, the annual monitoring report and local development scheme, all of which are mandatory, plus a suite of optional documents. Again, these plans are open to public scrutiny and are agreed upon between the local authority and regional planning bodies.

One of the most significant differences between the documents outlined here and the systems found in other parts of Europe is that of vagueness. For instance with the master planning in Germany, identically produced documents are produced at a fine-scale of geography, providing planning zones for the types of properties or structures that can be erected. In England, whilst it is possible to produce planning zones, most of the documents outline nothing more than applied policies and guidance, which can then give the local authority greater freedom in deciding on individual planning applications.

GIS in a Planning Context

Geographical information systems (GIS) provide some of the tools that can help planners in both assessing current plans and proposals, and predicting what outcomes they may produce. These can be through modeling or monitoring of systems, as in the example on brownfield land monitoring outlined in chapter "Monitoring the Effective and Efficient Use of Land: The English Approach" of this volume.

Using GIS to Define and Measure Town Centers

History of the Decline of the British Town Center

Town centers have an important role in British history and culture. Prior to the expansion of both private and mass-public transport, most of the population was unable to consider traveling long distances to their workplace or for shopping. Therefore the local center become highly significant to them, both in terms of commerce and social interaction (Thomas 1989), and became a focal locus for the local community. These loci covered (and still do today) a range of necessary activities: shopping, workplaces, leisure and a wide range of services such as banking, hairdressing, and travel agents.

With affordable and easily accessible public transport, and particularly with private cars, the public has been free to move around much more, opening up previously inaccessible locations. Overall urban densities have fallen (see chapter "Monitoring the Effective and Efficient Use of Land: The English Approach") as sprawl and the number of "edge cities" (Garreau 1991) have increased, all fuelled by the use of the family car (Schiller 2001).

Schiller (1986; 2001) argues that retailing in Britain has seen a shift toward decentralization from the high street from the 1980s onwards. He highlights three main waves of this decentralization, though it is asserted that we are currently living through a fourth.

Supermarkets lead the first wave, who, after seeing the increase in car use, saw the opportunity to offer what the customer was looking for: large free car parks located alongside the supermarket (Guy 1998). Suddenly there was an alternative to the prospect of driving in increasingly congested town centers, paying parking fees, and having to buy only what can be carried from the high street back to the car park. Instead people could buy as much as they desired and simply push their trolleys right to their car boot, load up and leave.

Following this opening move, the supermarkets were joined by the larger retail warehouses in the second wave. Typified by furniture and carpet retailers, electrical goods suppliers, and the DIY sector stores, the retail parks provided larger stores with more floorspace, allowing them to offer bulkier goods that smaller in-center stores could never stock (Guy 1994).

The third of Schiller's waves was heralded by Marks & Spencer's announcement in 1984 that they would start opening edge-of-center and out-of-town stores (Guy 1995). With this move, the traditional retailers selling less bulky goods who had remained solely on the high street saw an opportunity to get in on the trading taking place out-of-town. Brent Cross became the first regional shopping center in Britain, full of high street retailers taking advantage of out-of-town centers. Soon every mid-sized town in the country could find retail parks on their radial and main link roads (Lowe 2000). Many now offer more than just retail; restaurants, public houses, bowling alleys, and cinemas all help the parks to create their own "community," attempting to replicate the cultural life previously offered within the towns themselves (Gardner and Sheppard 1989).

The end-point (for the present) of the decentralization has been the ultimate in the consumer centered retail experience, with shopping, leisure, entertainment and food all under one closed roof: the shopping mall. The Gateshead MetroCentre was the first of these, founded on a Free Enterprise Zone near Newcastle in 1986. Since then many more have sprung up throughout the length and breadth of the country, in some cases controversially and arguably being built against the advice laid down in the planning guidance (Lowe 1998).

Internet retailing (or e-tailing) can be regarded as the fourth wave of retailing decentralization, albeit in a different form to that seen earlier. The rise of businesses such as Amazon.com demonstrated that the Internet could provide another channel for shopping: that of shopping from the comfort of one's own home. This is, of course, not new. Mail order shopping has been around for some time, but has been seen as only providing a limited market. The Internet has allowed many more people to access many more items more readily, and we are still seeing the upwards increase of e-tail.

As well as books and personal items, the e-tailing has become a success in the one retailing sector that dominates all others, groceries. Tesco first started the online grocery retail market, but was followed by Sainsbury's and Waitrose (Murphy 2002). It can be hypothesized that we have yet to see the interest in such services peak, and this may go on to have further impact on the traditional outlets on the high street.

Metrics for Town Center Monitoring

Retail decentralization has produced a number of downturns in the traditional town centers. There has been some evidence to show that the number of high street retailers has been in decline and this is particularly the case with the smaller independent stores. Between 1971 and 1979 the number of independents dropped by 35 per cent in comparison to a nine per cent fall in the number of multiples (i.e., retailers with multiple stores under the same brand name), shifting the relative share of the overall market away from independent retailers to the multiple retailers (Guy 1982). More recent work has suggested that small to medium sized centers have lost between 13 per cent and 50 per cent of their food trade (Thomas and Bromley 2002). All of this reduces the retail diversity of both local centers and the nation as a whole.

As well as some summary figures, as described above, there have been some particular examples highlighting the impacts. In Llanelli, Wales, the opening of a new out-of-town supermarket and the closing of the corresponding high street store led to a significant decline of town center retailing. With the loss of the supermarket, there was also a loss of "spin-off" shopping (where customers come mainly to visit one shop, but whilst there shop elsewhere as well). The economy of Llanelli continued to fall off until a second grocery retailer opened a new site within the town center in 1997, which sparked a renewed interest in the town (Thomas and Bromley 2002).

Whilst some of the impacts of retail decentralisation are known, much of the recent retail change has taken place within an information vacuum. From 1952, many statistics on the number of stores, employees and sales were produced from the government-run Census of Distribution (Sparks 1996). This information was useful for monitoring the health and sustainability of town centers (Thurstain-Goodwin and Batty 2001). The final one was taken in 1971 and since then the only data existing that could be used for such monitoring has been that produced by commercial organisations. However, from a public service point of view, these datasets are wholly unsuitable. Commercial ventures rarely publish how data collections are made and exactly what data they contain, are not usually consistently applied or collected across the country, and frequently are not collected for areas that they determine not to be commercially viable. Without such information, it has become impossible to assess the vitality and viability of town centers (Ravenscroft 2000; Ravenscroft et al. 2000), nor gain any understanding of the relative impacts of planning policies, particularly those coming from the central government planning guidance note PPG (later updated and published as PPS6: Town Centres).

Defining Town Centers With Statistical Modeling

To help deal with the problem of a lack of data, Her Majesty's Government commissioned a project to come up with a way of producing consistent, robust, and regular data. Being fundamentally a geographical problem, GIS was the most applicable tool to create such a model and output statistics.

An outline of the processes followed and a final model is provided here, but further details are covered elsewhere: see Department of the Environment Transport and the Regions (1998), Lloyd (2005), Lloyd et al. (2003), Office of the Deputy Prime Minister (2002a, 2002b) and Thurstain-Goodwin and Unwin (2000).

The first question that was asked was "what is a town center?" This highlights one of the criticisms of GIS in that whether the attempt to objectively define something that is nothing more than conceptual is a valid approach (Taylor 1995). We know when we are in a town center as we can rationalize the types of business found there and what it should look like. In this context the cultural indicators of what is and is not a town center are often immediately apparent to people with experiences of other centers and urban forms. However, defining this in a statistically robust model is much more complex.

The final model focused on three main components in that town centers are typified by:

a) A typical set of economies (i.e., retailing, some types of leisure, offices and some types of services; in particular not warehouses, factories, or heavy industries);
b) A high volume of commercial and industrial floorspace (again, though, not warehouses or factories); and
c) A wide diversity of types of activities – without which there is no way of separating a large out-of-time retail center (only containing retailing activities) from a town center (containing retail along with the other economies discussed above).

Existing government datasets provide all the information required to model these three components. The Office for National Statistics (ONS) produces the *Annual Business Inquiry*[1] containing estimates of turnover and the number of employees for each business unit in the country, broken down by their standard industrial codes (SIC[2]), which can be used as a proxy for the economy component. It can also be utilized to estimate the diversity of an area as the number of businesses with differing SICs will give an indication of the range of different jobs and industries represented.

Secondly, some of the Valuation Office Agency's (VOA) responsibilities are based around supporting the property-based taxes for England and Wales. This includes both Council Tax valuations for residential dwellings and Business Rates

---

[1]See www.statistics.gov.uk/abi/ for more detail

[2]See www.statistics.gov.uk/abi/sic92_desc.asp for more detail

for commercial and industrial properties. The latter is based on floorspace sizes, so the VOA, through a combination of their *Rating Support Application* and *Rating List* databases provide floorspace estimates and bulk land use types for each property.[3] This data set can then be used to provide estimates for the floorspace (or "intensity of use") component.

Using raster capabilities of a GIS package, a kernel density estimator can be used to smooth the postcode level data from ONS and VOA, and then map algebra is used to combine the three separate indicators into one final indicator called the Index of Town Centre Activity (ITCA). The detail of the process is described in full in the publications referenced at the start of this section.

Outputs and Dissemination from the Model

The ITCA can be regarded as a scale of "town centerdness." The higher the value the more associated the location is to a statistical definition of a town center. With knowledge gained from detailed local area studies it has been possible to create a set of rules for selecting a threshold value on the surface, dividing the ITCA into areas that are (or possibly are) town centers, and those that are not. In addition, a second model is produced that only uses the retail units in the ABI to define retail core areas of town centers.

A simple point-in-polygon can then be used to produce summed statistics on different types of employment and floorspace in each of the thresholded town centers. The exact outputs are outlined in Table 2. There is a longer-term aim to include the publication of turnover figures with the outputs, but at the time of publication the ABI turnover figures have not yet been proved to be robust enough for usage.

All the statistics for 1999 (London only), 2000 and 2002 (whole of England) have been published, with an aim of extending this series to run from 1998 onwards.

Table 2 Statistical outputs from the town centers model

Employment	Floorspace	Rateable value[4]
Commercial offices	Bulk retail	Bulk retail
Civic and public administration	Bulk office	Bulk office
Convenience retail	A1 use class	A1 use class
Comparison retail	A2 use class	A2 use class
Service retail	A3 use class	A3 use class
Arts, culture and entertainment		
Restaurants and licensed premises		

---

[3]See www.voa.gov.uk/business_rates/index.htm for more detail.

[4]See The Town and Country Planning (Use Classes) Order 1987 (www.opsi.gov.uk/si/si1987/Uksi_19870764_en_1.htm) for more details on the use classes.

As well as being put out in standard statistical tables, Geofutures have produced a Google Maps interface to allow the dissemination of boundaries and the associated statistics to a wider audience.[5]

The publication of the statistics and boundaries will allow users in both the public and private sector to gain an understanding of how each town center is changing over time – both in its economy and performance as well as its size and form. With such a tool it has become possible to assess both the short- and long-term impacts of policy and adjust them if necessary. Whilst it is still early days for this type of modeling and the outputs from this particular model, further work by Michael Bach and Geofutures on behalf of the British Council of Shopping Centres (Bach and Thurstain-Goodwin 2006) suggests that whilst significant development in out-of-town shopping centers continued unabated following the publication of PPG6: Town Centres in 1997, much of this was already in the pipeline. In more recent years development has shown to be more compliant with the planning guidance and therefore possibly has had less impact on town centers.

Reusing the Town Centers Methodology for a Housing Capacity Model

History of Housing Growth and Price Changes

Over recent years the British housing market has undergone enormous price changes. Housing affordability for England as a whole, as measured by the ratio between lower quartile earnings to lower quartile house prices, has risen from 3.65 in 1997 to 7.12 in 2006.[6] During the same period the mix-adjusted (which adjusts the prices of actual house sales in proportion to the types and sizes of houses that make up the whole stock) house price inflation rate has averaged 11 per cent annually, leaving the average house price for England in March 2007 at £214,424 against £129,973 in February 2002.[7]

Increasing house prices and lowering levels of affordability are causing many people and families difficulty across Britain. Whilst the proportion of owner-occupiers, at about 70 per cent of the households, has never been higher, an increasing number of households are being locked out of the market. The impact and causes of rising house prices are best discussed elsewhere – see, for instance, Bramley (2007) and Meen (2005) – however, one of the main causes is the supply of new housing.

The Barker Review of Housing Supply (Barker 2003, Barker 2004) and Government's Response (Office of the Deputy Prime Minister 2005) have highlighted that there have

---

[5]www.geofutures.com/online/towncentresfullscreen.html
[6]Data published by Communities and Local Government as part of their housing statistics Live Tables- http://www.communities.gov.uk/housing/housingresearch/housingstatistics/housingstatisticsby/
[7]Data published by Communities and Local Government as part of their monthly House Price Index Statistical Release - http://www.communities.gov.uk/housing/housingresearch/housingstatistics/housingstatisticsby/housingmarket/publications/house-price-index/

been a number of factors driving the current-day short supply. This has partly come about from an undersupply of new dwellings being built throughout the 1980s and 1990s, but in addition the reduction in the size of the average household and the corresponding increase in the number of one-person households have been an influence.

One of the main findings to come out of the Barker Review was that to stabilize house price inflation at 1.8 per cent there should be an increase of at least 70,000 additional houses on top of the circa 150,000 (at that point in time) added to the stock each year. The government's response to this was to set a target for there to be 200,000 additional dwellings a year in England (additional dwellings are defined as the number of new dwellings built minus the number of dwellings demolished, thereby not being the same as straightforward house building figures).

Planning Policy Statement 3: Housing[8] was drafted with this target in mind. Within the statement there is guidance on how this increased development should take place. This includes instructions that development should, where possible, be aimed at areas that are already developed. There is a priority list order of the type of land that should be used ahead of others. Filling in gaps within already existing urban areas (urban infill) takes priority over extending the margins of urban land, which in turn should take priority over constructing new communities. Thus there is a drive for densification, rather than new low-density developments. Density targets have existed for some time, with a current overall target of new dwellings being constructed at densities of at least 30 dwellings per hectare.

In addition brownfield (or previously-developed land) should always be used before greenfield sites. The rationale behind this is that it is preferable to reuse land that has already been used in the past but has now fallen into disuse than to build on the greenspace and land that might be required for other uses (for instance agriculture). Chapter "Monitoring the Effective and Efficient Use of Land: The English Approach", of this volume, discusses this in more detail.

Modeling Housing Capacity

To return to the opening description of government's role and responsibilities in planning, it should be clear that central government has a major problem: no matter how much guidance it produces, it only has limited ways of influencing house building at a local scale and understanding what capacity there might be on the ground.

One of the advantages of the town centers model outlined above is that the methodology is reusable for a number of situations. Early work has therefore started on adapting this technique to make an assessment of the suitability of land for housing. By using data such as existing dwelling densities, planning restrictions and environmental designations, previously-developed land densities, flood risk, natural resource capacities, employment rates and expected job growth, and socio-economic capacities of educational and health establishments, it is possible to produce an index of suitability.

---

[8]Published at http://www.communities.gov.uk/planningandbuilding/planning/planningpolicy guidance/planningpolicystatements/planningpolicystatements/pps3/

It must be stressed that such a methodology, if successful, can only be used for gaining a broad understanding of the potential capacity. How suitable a location is for housing is a question that should only be properly answered from a local perspective. To do otherwise would fall foul to the criticisms levelled at GIS since its first use (Schuurman 2000; Taylor and Overton 1991). However, such a tool can give high level guidance to areas of the country that are (empirically) suitable for further development, as well as providing the center a way of supporting local level planning.

Conclusions

GIS and the Planning System: Providing Tools to Help or Hinder?

GIS can offer a number of ways of supporting ongoing work and providing such detailed geographical analysis that can be critical to fully understanding a problem. It can input into the creation of evidence-based policy and produce a monitoring framework that can demonstrate how successful the relevant policies are, allowing early detection of errors, mistakes, and failings. However, it needs to be recognized that there are limits, and that it can be used in such a way to actually hinder, rather than help.

In the case of the town centers work, controlling the use of a "statistical boundary" in an environment where policy boundaries are common can be hard. Where practitioners (at all levels of government and outside) are used to dealing with fixed boundaries drawn up on the ground specifically for a local area, a boundary that is statistically derived by a generic model and that can vary over time brings new ways of thinking. Equally, to rely wholly on a top-level model of housing capacity when there are so many local level influences and controls would be a mistake. The difficulty here is not necessarily defining the correct model, but rather knowing what to apply it to, and knowing when it becomes unsuitable.

References

Bach M and Thurstain-Goodwin M (2006) Future of retail property: in town or out of town? Geofutures Ltd and British Council of Shopping Centres
Barker K (2003) Review of housing supply: interim report - analysis. HMSO, Norwich
Barker K (2004) Review of housing supply: final report - recommendations. HMSO, Norwich
Bramley G (2007) The sudden rediscovery of housing supply as a key policy challenge. Housing Studies 22: 221–241
Department of the Environment Transport and the Regions (1998) Town centres: defining boundaries for statistical monitoring feasibility study. The Stationery Office, London

Gardner C and Sheppard J (1989) Consuming passion: the rise of retail culture. Unwin Hyman, London

Garreau J (1991) Edge city: life on the new frontier. Doubleday, New York

Guy CM (1982) 'Push-button shopping' and retail development. Department of Town Planning, UWIST, Cardiff

Guy CM (1994) The retail development process: location, property, and planning. Routledge, London

Guy CM (1995) Retail store development at the margin. J. of Retailing and Customer Services 2: 25–32

Guy CM (1998) Off centre retailing in the UK: prospects for the future and the implications for town centres. Built Environment 24. 16 30

Lloyd DA (2005) Uncertainty in town centre definition. PhD thesis, UCL, London

Lloyd DA, Haklay M, Thurstain-Goodwin M and Tobón C (2003) Visualising spatial data structure in urban data. In: Longley PA and Batty M (eds) Advanced spatial analysis: the CASA book of GIS. ESRI Press, Redlands, CA, pp 267–288

Lowe MS (1998) The Merry Hill regional shopping centre controversy: PPG6 and new urban geographies. Built Environment 24: 57–69

Lowe MS (2000) Britain's regional shopping centres: New urban forms? Urban Studies 37: 261–274

Meen G (2005) On the economics of the barker review of housing supply. Housing Studies 20: 949–971

Murphy A (2002) The emergence of online food retailing: a stakeholder perspective. Tijdschrift voor economische en sociale geografie 93: 47–61

Office of the Deputy Prime Minister (2002a) Producing boundaries and statistics for town centres: London pilot study summary report. The Stationery Office, London

Office of the Deputy Prime Minister (2002b) Producing boundaries and statistics for town centres: London pilot study technical report. The Stationery Office, London

Office of the Deputy Prime Minister (2005) Government response to Kate Barker's review of housing supply. HMSO, London

Ravenscroft N (2000) The vitality and viability of town centres. Urban Studies 37: 2533–2549

Ravenscroft N, Reeves J and Rowley M (2000) Leisure, property, and the viability of town centres. Environment and Planning A 32: 1359–1374

Schiller R (1986) The coming of the third wave. Estates Gazette 279: 648–651

Schiller R (2001) The dynamics of property location. Spon, London

Schuurman N (2000) Trouble in the heartland: GIS and its critics in the 1990s. Progress in Human Geography 24: 569–590

Sparks L (1996) The Census of Distribution: 25 years in the dark. Area 28: 89–95

Taylor PJ (1995) Geographic information systems and geography. In: Pickles J (ed) Ground truth: the social implications of geographic information systems. Guilford Press, London, pp 51–67

Taylor PJ and Overton M (1991) Further thoughts on geography and GIS – a preemptive strike. Environment and Planning A 23: 1087–1090

Thomas CJ (1989) Retail change in Greater Swansea: evolution or revolution. Geography 74: 201–213

Thomas CJ and Bromley RDF (2002) The changing competitive relationship between small town centres and out-of-town retailing: town revival in South Wales. Urban Studies 39: 791–817

Thurstain-Goodwin M and Unwin DJ (2000) Defining and delineating the central areas of towns for statistical monitoring using continuous surface representations. Transactions in GIS 4: 305–317

Thurstain-Goodwin M and Batty M (2001) The sustainable town centre. In: Layard A, Davoudi S and Batty S (eds) Planning for a sustainable future. Spon, London, pp 253–268

Monitoring the Effective and Efficient Use of Land: The English Approach

Robin Ganser

Abstract In England quantified targets for the reuse of brownfields and of existing buildings place particular requirements on the planning system as part of the overall strategy to make effective use of natural resources. Achieving these targets is not free of obstacles and potential conflicts. Therefore monitoring of policy implementation is a necessity in the planning system.

In addition challenging targets for increased housing provision have been introduced. In view of this the dichotomy of adequate housing-land supply without compromising environmental quality must be addressed. As a consequence monitoring will play an increasingly important role.

This paper explores how these challenges can be met.

Brownfield Redevelopment: Necessity of Monitoring

Sustainable urban development is embodied in planning and environmental law in England. In this context a national target has been set to deliver 60% of all new housing on brownfields, respectively previously developed land (PDL), and through conversions by 2008 (hereafter "60% target") (DETR 2000a).[1]

---

[1] In this context it is interesting to note that PPS3 changes the terminology "previously developed land" to "brownfield land." The latter so far lacked a uniform classification at the national level. The following broad definition is applied to both terms now (ODPM 2005a): "Previously developed land is that which is or was occupied by a permanent structure (excluding agricultural or forestry buildings), and associated fixed surface infrastructure. The definition covers the curtilage of the developed land." Brownfields by this definition are not confined to derelict and potentially contaminated sites-instead they are understood to cover all previously developed land.

R. Ganser
Department of Planning, Oxford Brookes University, Headington Campus,
Gipsy Lane, Oxford, OX3 0BP, UK, E-mail: ganser@brookes.ac.uk

G. Steinebach et al. (eds.), *Visualizing Sustainable Planning*,
© Springer-Verlag Berlin Heidelberg 2009

So far England and Germany are the only countries in the EU that have integrated quantified brownfield targets in their national sustainability strategies and policy guidance (Ganser 2005). In both countries the motivations and overarching objectives that led to the introduction of the targets are similar. In this context the extent of PDL resulting from economic restructuring and demographic changes, counter-urbanization with population decline in large cities, and negative impacts on neighboring sites arising from derelict land have to be mentioned. Further, the rapid development of greenfield sites is linked to numerous negative ecologic effects (Ganser 2005).

The planning system has an important role to play in achieving the aforementioned targets through planning and development control procedures. Taking into consideration the general need for monitoring in the planning system, and more specifically the necessity to monitor the implementation of quantified brownfield targets, two fundamental questions arise:

1. Are the existing monitoring requirements and mechanisms geared for the tasks at hand including:

   Adequate measuring of brownfield targets – on a regular basis and over the entire planning horizon (60% target to be achieved by 2008)
   Complete coverage of the two overarching objectives – i.e., to further urban regeneration and minimise greenfield development through reuse of PDL
   Detection of potential conflicts – i.e., reduced greenfield development and adequate housing supply?

2. Which improvements or challenges are to be expected from the proposed changes to monitoring in the scope of the planning reforms?

Key Monitoring Principles and Statutory Requirements

The Tiered Indicator System in England – Strategic and Operative Level

At the national level the planning acts as well as planning policy and guidance set out the statutory monitoring requirements of Regional Spatial Strategies (RSS), which will be replaced by Integrated Regional Strategies, and Local Development Frameworks (LDFs). The monitoring duties placed on the planning bodies and authorities respectively can be differentiated as follows:

- Monitoring of general trends and conditions that are of environmental and spatial relevance. These monitoring results can inform the survey and review process of plans and programs as well as compulsory strategic environmental assessments and sustainability appraisals.

- Monitoring of quantified targets and overarching objectives in terms of progress toward these goals including an evaluation of the achievement of targets and the necessity to adapt these.
- Monitoring of the effects going hand-in-hand with the achievement of the afore-mentioned targets.

The three monitoring tasks above thereby form a "strategic level" of monitoring that is primarily concerned with the evaluation of achievements.

Further to this an "operative level" can be distinguished that focuses on the implementation of quantified targets and linked indicators in planning documents or strategies and through development control. This includes:

- Making sure that all necessary targets and indicators are in place.
- Controlling the formal planning processes – e.g., dynamic determination of plan reviews according to monitoring results.
- Support of the decision-making process through monitoring results – e.g., providing the framework for discretionary decision making in the scope of planning and application procedures.

The English monitoring system is therefore not confined to informing policy makers and the public of the real world effects of policy implementation. Its role in the planning system is far more comprehensive, as it can provide an active means of controlling and managing planning processes and urban development if used correctly.

Double Role of Monitoring

Monitoring principles and results are documented in so-called annual monitoring reports (AMRs), which have to be prepared for the Secretary of State (ODPM 2004b). In this context the AMR fulfills a double role: On the one hand it is the appropriate instrument for planning bodies and authorities to continuously gauge the progress of strategy or plan delivery, on the other hand it informs the Secretary of State – thereby providing an indication on whether further control and default powers need to be exercised to achieve national targets.

In addition, so – called Best Value Performance Indicators (BVPIs) have been introduced to evaluate the performance of local authorities (LAs). Some BVPIs are focused on procedural performance only (e.g., period of time taken for deci-sions on planning applications); others are identical or linked to material sus-tainability targets (e.g., percentage of homes built on previously developed land). The latter can provide an additional incentive for LAs to achieve the 60% target.

The following diagram gives an overview of the fundamental principles that characterize the monitoring system.

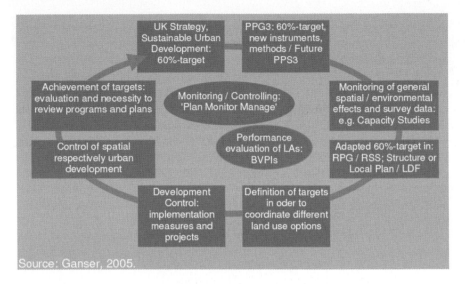

Fig. 1 Core policies and principles of English monitoring system

Structure and Methodology of the National Monitoring System in England

Since experience with the recently amended national monitoring system is very limited, the following analysis initially focuses on the system that was operational before 2005 (DEFRA 2005). In this context it is important to note that the new system does not introduce big structural changes, but comprises less indicators than its predecessor, which was established in 1999 (DETR 1999c). The latter consists of 147 indicators, including a set of 15 highly aggregated headline indicators, which are used to monitor central aspects of sustainable development (ibid). Headline indicator H14 measures "new homes built on PDL" (ibid). Further to this indicators that cover the "net loss of soils to development" (S1) and "vacant land and properties and derelict land" (K1) are used to monitor brownfield as well as greenfield development (ibid).

The monitoring system with its tiered structure covering the national, regional, and local level resembles a mirror image of the planning system. The basic principle in this context – shown in Fig. 2 – is the provision of vertical compatibility in order to allow efficient data collection and analysis through clearly defined responsibilities at different levels, thereby minimizing unnecessary duplication of work.

Monitoring in the planning system therefore requires a parallel top down and bottom up process of defining targets, linking indicators, and collecting as well as analyzing necessary information. The starting point for the definition of indicators is the set of national (headline) indicators. At the regional level additional indicators may be defined to adhere to the specific regional context. The sub-regional and local monitoring systems have to be compatible with the higher tiers of planning,

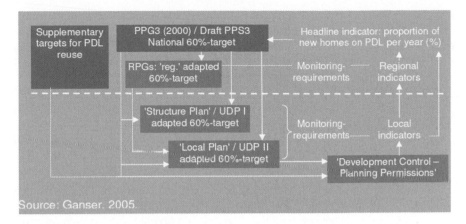

Fig. 2 Tiered structure of the English monitoring system for brownfield reuse

and can again be supplemented by location specific indicator sets. Monitoring results from the local level are subsequently fed upwards and provide a base for regional and national monitoring.

Due to the tiered structure compatibility between the different monitoring levels is crucial for the quality of monitoring results.

National Monitoring Framework and Central Databases

Adequacy of National Standard Indicators

The analysis of the English monitoring system (pre-2005) shows that the achievement of the 60% target is adequately addressed at the national level through headline indicator H14. However, the overarching objective to further urban regeneration is not sufficiently covered by indicators H14 and K1. The latter merely depicts brownfields that may be appropriate for reuse. It is not possible to tell from this particular indicator if an increase in brownfields is primarily due to new sites that fall derelict, or due to stagnating reuse. Information on the quantitative contribution of PDL reuse to overall development (all uses) is only available up to 1998 (English Partnerships 2003).

However the data from the National Land Use Database (NLUD) suggests that the entire stock of PDL has increased from 57,000 ha in 1998 to some 67,000 ha in 2002. Evidence from research implies that a "hardcore" of PDL exists that has remained unused for over 10 years, while new "easy-to-develop" brownfield sites are brought back into use earlier, thereby revolving around this hardcore (ibid). The regional distribution of hardcore sites is uneven. Particularly the North West shows a concentration of these sites (4,000 ha of 11,000 ha total). This in turn leads to the assumption that the risk of running out of the easy-to-develop sites is higher there than, for example, in the South East, where only 1,000 ha of 10,000 ha in total belong to the hardcore (ibid). The pace of brownfield redevelopment for housing from 1992 until 2001 in the North West (400ha/a) and in the South East (500 ha/a) underpins this assumption (ODPM 2003a).

Information is not available on qualitative aspects such as locational advantages at the micro scale or possible synergies with adjacent development etc. As a result the monitoring arrangements at the national level do not allow precise conclusions on the contribution of the 60% target to further the process of regeneration.

Until recently the overarching objective to reduce greenfield development was covered by the national indicator S1, which measures the absolute loss of soils to development (all land uses). The analysis of respective time series can highlight the general trend of development: The number of completions fell drastically to 49,600 units per year from 1999 to 2002, compared to 62,900 units per year during the preceding four year period (ODPM 2004a). Parallel to this the total greenfield area developed for the first time decreased considerably (ODPM 2003a).

Due to the highly aggregated nature of the national indicators, unwanted negative effects cannot be detected directly through the monitoring system. In particular the potential conflict between reduced greenfield development and adequate supply of housing requires more information on the following parameters: the total allocations for housing in development plans that have not yet been taken up by development compared to the annual targets for housing completions. If the completions cannot be accommodated through the sum of outstanding planning permissions, windfalls, and aforementioned allocations, a conflict arises (Ganser 2005). If on the other hand the annual completion rate lags considerably behind the issued planning permissions – economic reasons for holding back development rather than impediments through the planning system can be assumed.

Further to this point, the national indicator set does not measure the implementation of the 60% target and linked objectives, such as, sufficiently imbedding the national targets in the local development plans and adequately linking implementation measures and meaningful indicators. Therefore hindrances posed by the planning system in this respect cannot be detected immediately.

Data Bases and Data Collection

Four additional monitoring deficits at the national level are linked to the method of data collection. The Land Use Change Statistics (LUCS), which is the central source for indicators H14 and S1, usually involves a considerable time lag between the completion of new development and data collection through analysis of aerial photography and site visits. Due to this methodology the change of use in existing buildings that does not involve physical alterations of the built structure is difficult to detect. Some periods (e.g., 1999) had to be complemented by assumptions as base data showed gaps (Ganser 2005).

LUCS is focused on status quo information – it does not include information on planned land-use allocations. It comprises a post-evaluation of land-use development. As a consequence, additional information has to be obtained in order to judge the future ability to meet the 60% target and adequate housing provision.

The aforementioned NLUD is an important source of information for the achievement of the 60% target in the future. NLUD is based on returns from Local Authorities (DETR 2000b). As there is no legal obligation for these returns they are conducted on a voluntary basis. Data is collected by the local authorities and transferred to ODPM where it is fed into NLUD. The legal ownership of the data remains with the LAs. This in turn means that the information on developable brownfields is only accessible through applications to individual authorities, which can cause a hindrance for commercial developers as well as for private individuals who are looking for sites with specific characteristics rather than sites at a certain location.

NLUD offers a broad range of spatial information (ibid):

- LA and location
- size of site
- type of site (e.g., industrial brownfield)
- relevant physical and planning restrictions
- plan allocations and planning permissions
- most suitable use or mix of uses
- appropriate densities
- potential yield in terms of number of dwellings
- availability of sites, owner information and
- potential financial contributions (e.g., from EP or RDAs).

Site types are categorized from A to E comprising:

A) PDL that is now vacant
B) vacant buildings
C) derelict land and buildings
D) land or buildings currently in use and allocated in development plan and/or having planning permission
E) land or buildings currently in use where it is known there is potential for redevelopment but sites do not have plan allocation or permission.

It is clear from this list that NLUD is set up to be forward looking, as it includes sites that are underused or show future development potential. This can help to close the gap between completions on allocated sites and windfalls thereby providing greater certainty for planners and developers.

Unfortunately NLUD just like LUCS is afflicted with several deficits. Probably most problematic are variations in data collection returns leading to results that are not directly comparable – as a consequence precise timelines from 1999 until today are not yet available. This in turn leads to gaps in the understanding of the lifecycle of PDL and its reuse.

Further to this NLUD does not comprise information on the environmental quality of the site (fauna, flora etc.). Questions regarding potential reclamation and reuse for ecological purposes – like stepping stone habitats within a biotope network – and other soft end uses therefore remain unanswered.

Additionally NLUD does not provide information on financial viability of site development, other than evidence of restrictions, including contamination. For instance, it is therefore possible that there are sites allocated in the development plan that are not viable in the foreseeable future.

In addition to this, minor methodical deficits occur. For instance there is a certain ambiguity concerning density parameters for mixed-use schemes with an industrial and housing element, as the former is measured on the base of floor-space and the latter on dwellings per hectare.

In parallel to NLUD the so-called "Register of Surplus Public Sector Land" has been established through EP by order of ODPM with the aim to further a swift reuse of publicly owned sites (www.englishpartnerships.co.uk 2004). It is questionable why these records are provided in a separate database and are not included in NLUD, as this frustrates the potential one stop shop for information on developable brownfields at national level.

The above deficits in combination compromise the precision of indicator-based monitoring at the national level. This in turn reduces the value of the monitoring results as a basis for decision-making to the effect that the consequent review and amendments of targets and linked indicators can be encumbered.

Main Shortcomings of Monitoring in Planning Practice

An analysis of planning practice at the regional and local level shows that the monitoring deficits discovered at the national level are only partially made up for by subordinate tiers of planning. Fig. 3 provides an overview of the main shortcomings that were detected in English planning practice.

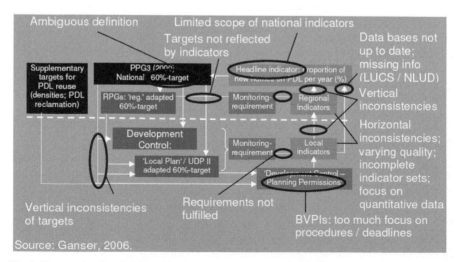

Fig. 3 Key monitoring deficits:Affected elements in the monitoring system

The deficits shown in Fig. 3 do not mean that results from the entire system are meaningless, but due to the somewhat fragmented nature of monitoring results a higher degree of interpretation and sometimes additional data collection is necessary. Only under this premise can existing monitoring make a positive contribution to spatial planning and brownfield reuse in particular.

In this context the second question raised at the beginning needs to be explored. The proposed changes to monitoring practice and their potential to cure the above problems are therefore examined subsequently.

New Requirements for Monitoring and Managing Land Supply

Rationale for New Demands on the Planning System and Key Instruments

Land supply constraints were named as a key hindrance for housing delivery in the Barker Review (Barker 2003). This is singled out in the "Planning for Housing" consultation paper and in PPS3, which discuss how planning should deliver housing at the local level and offer proposals for new delivery mechanisms (ODPM 2005a, ODPM 2005d). In this context the following short diversion is necessary: it needs to be considered that the phrasing of "housing delivery" might give rise to false expectations with regard to the role of the planning system. Given the fact that the planning system is not endowed with sufficient funds that directly serve the purpose of house building and infrastructure provision, the LAs will not be able to take on housing delivery single-handedly, as this clearly requires a multi-stakeholder approach. Great care needs to be exercised in order to prevent spatial planning from being a scapegoat for not meeting housing completion targets. Clearly the focus has to be on identifying and allocating adequate sites in the right locations at the right time. The key role of the planning system – more specifically of the LAs – therefore is to secure adequate land supply in order to facilitate residential and mixed-use development.

The following planning instruments and methodological elements are at the heart of the new proposals: RSS should, in the future, set out the level of housing provision for the region for at least 15 years, the level of housing provision for the plan period for each identified sub-regional housing market area, and for each LA within it (ODPM 2005a, Sect. 34). Additionally RSS should define the region's brownfield target and density target, and/or the region's density range, respectively (ODPM 2005a, Sect. 42, 45). The LDF subsequently should allocate sufficient land and buildings for housing, or mixed-use development, to deliver the first five years of housing provision – taking into account a windfall allowance only where it is not possible to allocate sufficient land (ibid, Sect. 59). Further, for the next 11–15 years of housing supply should be allocated to this land (ibid, Sect. 55). Where it is not possible to allocate specific land, broad areas of land for future growth should be indicated in the core strategy.

If the housing targets are not met the following actions are envisaged: identifying and removing barriers to the delivery of sites allocated for release in the five-year land supply; review the phasing of existing allocations; and/or allocate new developable land for housing, including the allocation of specific sites within the broad areas indicated for development in the core strategy through an update to the site allocation development plan document. In this context a remarkable change from PPG3 to PPS3 is noticeable. Whereas the former placed the main emphasis on the provision of housing land at the most sustainable location on the most sustainable sites, the latter is more directly focused on sufficient land supply in order to meet market demand.

The new requirement placed on LAs to regularly roll forward housing land allocations – although with the priority to bring forward brownfield development – could mark a return to the "predict and provide" concept. So far it is unclear how a potential conflict between the local brownfield target and the housing target – both part of the LDF – are to be resolved in the scope of the plan-monitor-manage approach. In order to prevent adverse effects on the achievement of brownfield targets and the linked objectives, monitoring will play an increasingly important role, which is explored below.

New Monitoring Requirements in the Scope of the Planning Reforms

In 2005, at the national level the set of headline and additional indicators was replaced by a new set of "framework indicators" and "supporting indicators." As regards monitoring of greenfield and brownfield development, the land-recycling indicator (percentage of dwellings on brownfields and through conversions) can only be found in the latter category (www.sustainable-development.gov.uk 2005). This also indicates that the political emphasis has somewhat shifted from housing provision at the most sustainable location to meeting housing demand. Complementary to this a new indicator covering "all new development (not just residential) on previously developed land" has been introduced.

Unfortunately indicator S1 "loss of soils" to development is no longer part of the national indicator set. The new "land use" indicator merely depicts the percentages of land area taken up by specific land uses, including "urban land / land not otherwise specified" (ibid). In terms of absolute greenfield development this indicator is less clear than its predecessor.

Both the Planning and Compulsory Purchase Act 2004 and PPS3 Housing underline the importance of monitoring in the planning system, and more specifically of the AMRs. Planning for Housing states the following main reasons for monitoring:

- To enable local authorities to understand the reasons why they may be failing to deliver their planning housing provision.

- To enable local authorities to prevent inappropriate delays in land supply by rolling forward their five-year supply as it is developed (ODPM 2005d, Pl&CP Act 2004).

Neither of these reasons relates to the reuse of brownfields or to the implementation of the sequential approach (priority of brownfield development) in this context. This rather narrow focus is also reflected in the monitoring requirements that are directly set out in PPS3, which include monitoring of housing permissions and completions, whether on previously-developed land or greenfield in both urban and rural areas (ODPM 2005a, Sect. 76). These requirements are complemented by separate good practice guidance, which is more relevant for monitoring housing development on brownfields (ODPM 2005e, 2005g).

At the regional level the following standard indicators for monitoring RSS relate to brownfield development (ODPM 2005f):

- percentages of business development by type, which is on previously developed land: by local authority area
- comparing (in percentage terms) the amount of completed land (sq. meters gross floorspace) developed for employment upon previously developed land against the total completed employment gross floorspace
- percentage of new and converted dwellings on previously developed land
- comparing (in percentage terms) the number of completed dwellings (gross), and conversions of existing dwellings provided on previously developed land against total completed dwellings (gross)

A comparison between old and new monitoring requirements with relevance for housing development on brownfields at the local level shows that PPS3 comprises only a very limited number of indicators and will therefore be heavily reliant on additional guidance. Its predecessor in contrast provided a complete outline of indicators that were merely complemented, respectively detailed, by practice guidance. In addition to the above, guidance on Sustainability Appraisals of RSS and LDF suggests the following indicators:

- proportion of land stock that is derelict
- land covered by restoration and aftercare conditions
- vacant land and properties and derelict land

An examination of the "operative level" of monitoring reveals that most of the new requirements are merely clarifications that have already been implemented in the scope of good practice.

The AMR is now a statutory requirement at the local level. In the scope of a so-called housing trajectory, LPAs should assess any future shortfall or surplus of housing over the plan period by comparing this to planned building rates (ODPM 2004b). This will have to be updated regularly.

As part of the AMR, local planning authorities should report on progress against the housing and previously-developed land trajectories. In this context it is possible to define acceptable ranges of target achievements. Further to this they should set out the actions to be undertaken where actual performance does not meet the trajectories (ODPM 2005a, Sect. 76).

Additionally, the delivery performance should be considered in the context of the housing market area and the region as a whole (ibid).

Resolving Existing Problems and Potential New Deficits – The Need for Further Development and Improvements

The following compilation highlights which monitoring deficits have been cured as a consequence of the new requirements introduced above, and where there is need for further development and improvements.

Potential Displacement Effects

At the national and regional levels the deficit of not being able to spot displacement effects, whereby the housing priority on brownfields might push other land uses to less suitable sites, is partially remedied by the new indicators for business development on brownfields compared to greenfield up-take. This however requires cross comparisons and some degree of interpretation, e.g., monitoring results for housing and other land uses have to be analyzed jointly. In this context it appears to be necessary to provide "interpretation rules" to make sure, amongst other things, that adequate methodologies are used.

Ambiguous Definition of 60% Target and Linked Indicator

The (communication) deficit due to the fact that the monitoring of the 60% target does not allow an immediate judgement of progress towards the overarching objectives of reduced greenfield development in absolute terms and furthering regeneration is not remedied. As far as the former is concerned the new "land use indicator" covers only developed land in percentage terms, and therefore offers less information than the old indicator "net loss of soils."

To improve this situation the following indicators should be measured at all levels of the planning system: "loss of soil to development (by development type, measured in ha per annum)," and "absolute reclamation of brownfield land (by development type measured in ha per annum)."

Conflicts Between 60% Target and Housing Targets

Conflicts between reduced greenfield development and sufficient land supply can now be more easily detected at the regional and local levels. Particularly

LDF monitoring aims to provide a fine grid of indicators picking up housing permissions, starts and completions on land allocated for housing, and on windfalls. Existing and projected allocations are also monitored. There is however no indicator that covers the absolute amount of greenfields changed to developed use. Also, the number of outstanding planning permissions is not covered by the standard indicators. Under the proposed system this can mean that a delay between permissions and completions can lead to increased housing land allocations in the LDF. This in turn can defeat the objective of reduced greenfield development.

In addition to the indicators proposed under section 2 above, the outstanding planning permissions should therefore be monitored.

Unclear Timescale for Measuring

With the introduction of statutory annual monitoring reports the timescale for monitoring has been clarified.

Development of Non-Residential Land Uses not Linked to Brownfield Indicators

The new standard indicators at the regional level covering business development on brownfields in percentage terms partially remedy this deficit. A problem still remaining is the lack of monitoring in absolute terms: the absolute area of brownfields and greenfields developed as employment land should be included as indicators.

Lack of Data on Absolute Greenfield and Brownfield Development – Overarching Objectives of Reduced Greenfield Development and Furthering Urban Regeneration not Adequately Addressed

The monitoring deficit across all tiers of planning of the absolute amount of greenfield and brownfield development is partially remedied at the national level. The LUCS22 now provides data on the absolute change up to 2004 (p 18).

It should be possible to keep records more up to date with the new requirements to monitor planning permissions and completions at the local level. This however goes hand in hand with an increased need for vertical coordination between the planning tiers.

Ideally disaggregated data for all types of development – including leisure uses and transport infrastructure – should be collected. Only the interpretation of the overall picture, including the aforementioned land uses, which could thus be provided, would allow an adequate review and targeted changes to planning policies at all levels.

Monitoring of Planning Permissions (Housing) with Gaps

The new requirement on RSS and LDF monitoring reports to provide data on planning permissions – particularly for comparison with completions and allocated land – can remedy this problem. However there are no indicators covering planning permissions and completions for employment uses. Adequate indicators covering these two aspects should be introduced to rectify this deficit.

Not all Regional Targets and Indicators Reflected in Structure and Local Plans

With the introduction of RSS (including sub-regional strategies) and LDFs there is a more direct link between strategic policies, respectively targets, at the regional level and local policies. Additionally, the RSS now forms a statutory part of the development plan. Both amendments can contribute to better coordination and can prevent loss of material contents of planning documents. On the other hand, the flexibility of LDFs encourages partial reviews confined to specific local development documents (LDDs). It will therefore be a key task to ensure consistency throughout all the LDDs, which will also be reflected in increased demands on consistent monitoring.

Integration of National Targets in Regional and Local Planning Documents not Monitored – Vertical Inconsistencies

The insufficient incorporation of the 60% target in regional and local planning documents has been a problem in the past. It was partially based on the belief that national policy, like PPG3, is of direct relevance in the planning application process, and therefore does not need to be incorporated in local policies. This approach, however, does not reflect the need for variation according to regional and local context. The new monitoring requirements do not remedy this situation.

In this context overall implementation could be improved by introducing an indicator that covers the incorporation of brownfield targets in all statutory planning documents. This should, of course, include the detailed area action plans.

In order to resolve the shortcomings of vertical inconsistencies in the future, a set of standard indicators for all LDF areas within the region should be devised alongside national guidance. In this context there is a particular need for intra- and inter-regional coordination across authority boundaries.

Lack of Qualitative Indicators

There still is a lack of qualitative indicators such as the "suitability of land uses" on PDL, and the quality of residential and business development on brownfields that achieve the aspired high densities. This necessity arises to achieve a more balanced indicator set.

The introduction of an indicator covering "residential quality," e.g., according to "Building for Life" standards, would be a step in the right direction (www.buildingforlife.org 2006).

In addition to the above, the following new monitoring problems can arise:

Definition of Accessibility Too Narrow

The new monitoring field of public transport time, especially with regard to social infrastructure etc., is too narrowly defined. Unfortunately the more important questions of accessibility of public transport stations or stops (e.g., 400 m radius), and public transport quality (e.g., 10 min. frequency), which are decisive for public transport use are not addressed so far. Adequate indicators covering these criteria could improve this situation.

Ambiguous Density Monitoring

Another problem can arise in that the proposed density indicators and the indicative national target of 30 dph will be difficult to implement in the context of fine-grained mixed-use development incorporating different uses in the same building (ODPM 2005a, 2005g).

At the same time monitoring of densities will be of growing importance in the current period of transition to the new planning system. In this context the new requirement to adhere to the indicative national minimum density of 30 dph as long as no local targets have been defined may lead to a situation of inadequate densities particularly at sensitive rural locations.

It is equally important to define density indicators for employment uses, particularly in order to further the achievement of the overall target of reduced greenfield development.

Conclusions and Necessary Advances in Monitoring Brownfield Development

The structure and key principles of the English monitoring system make it an important instrument in strategic and operative terms. The exemplary analysis of current experience in planning practice highlights that some existing deficits could be eradicated through the introduction of the new monitoring requirements. On the other hand considerable need and room for improvement prevails. In particular, the uncertainty over progress toward the overarching objectives to reduce greenfield development while at the same time furthering urban regeneration through increased brownfield reuse has to be overcome. In the context of a new emphasis on meeting housing needs and demands this aspect will be of increased significance. In parallel to the presumably higher pressure on different locations and sites for (high density) housing development, the question of introducing qualitative indicators is more important than ever before.

Another issue of key importance is the necessity of interpretation of monitoring results, since, due to considerations of resource efficiency and transparency, only a limited number of indicators can be measured. Potential conflicts in particular can only be detected in the scope of a joint analysis including all relevant indicators. Concise guidance on the interpretation and comparative analysis of different indicators will be the key to successful monitoring of brownfield development.

After the successful implementation of the 60% target – not least through monitoring and controlling mechanisms – the major challenge for the future is to provide a more comprehensive picture of spatial development. The analysis and the proposed amendments above indicate possible improvements in this respect. These in turn could provide the foundations for the achievement of realistic brownfield targets covering all land uses, and in the long-term lead to a complete recycling loop of brownfields.

Last but not least it should be emphasized that, with respect to resource efficiency and consistent quality, the alignment of the new statutory requirements on SEA and on monitoring referring to all levels of spatial planning provides a challenge and at the same time a great opportunity to overcome the above deficits and, moreover, to consolidate the entire monitoring system.

References

Barker K (2003) Review of housing supply, securing our future housing needs, interim report. TSO, London

Bundesamt fuer Bauwesen und Raumordnung [Federal Office for Building and Regional Planning] (2003) Future takes place. Bonn, p 48

CLG (Communities and Local Government) (2006) Planning policy statement 3: housing, London. CLG Publications, Wetherby

CLG (Communities and Local Government) Department for the Environment Food and Rural Affairs, Department of Trade and Industry, Department for Transport, (2007) Planning for a sustainable future white paper TSO, London

CLG (Communities and Local Government) (2007a) Land use change in England: residential development to 2006, LUCS 22. Communities and Local Government Publications, Wetherby

Cullingworth B, Nadin V (2003) Town and country planning in the UK. Routledge, London

Dartford Borough Council (2004) First monitoring report 2004. Dartford

DEFRA (Department of the Environment, Food and Rural Affairs) (2005) Securing the future delivering UK sustainable development strategy. The UK government sustainable development strategy. TSO, London

DETR (Department of the Environment, Transport and the Regions) (1999a) A better quality of life: a strategy for sustainable development in the UK. TSO, London

DETR (Department of the Environment, Transport and the Regions) (1999b) Planning policy guidance note 12: development plans. TSO, London

DETR (Department of the Environment, Transport and the Regions) (1999c) Quality of life counts, indicator report. TSO, London

DETR (Department of the Environment, Transport and the Regions) (2000a) Planning policy guidance note 3: housing. TSO, London

DETR (Department of the Environment, Transport and the Regions) (2000b) National land use database, data specification. TSO, London.

English Partnerships (2003) Towards a national brownfield strategy, research findings for the Deputy Prime Minister

Ganser R (2005) Quantified targets for reduced greenfield development in the English planning system: a model for regional and local planning in Germany? Kaiserslautern Press, Kaiserslautern, TU

GONW (Government Office for the North West) (2003) Regional planning guidance for the North West region, RPG13. TSO, London

GOSE (Government Office for the South East) (2001) Regional planning guidance for the South East, RPG9, 2001. TSO, London

KCC (Kent County Council and Medway Council) (2003) Kent and Medway draft-structure plan

NWRA (North West Regional Assembly) (2003) Regional monitoring report 2003

ODPM (Office of the Deputy Prime Minister) (2002) Monitoring regional planning guidance. TSO, London

ODPM (Office of the Deputy Prime Minister) (2003a) Land use change in England: LUCS 18. ODPM Publications, Wetherby

ODPM (Office of the Deputy Prime Minister) (2003b) Land use change in England: LUCS 18A, land use change in England to 2002: additional tables. ODPM Publications, Wetherby

ODPM (Office of the Deputy Prime Minister) (2004a) Land use change in England: LUCS 19. ODPM Publications, Wetherby

ODPM (Office of the Deputy Prime Minister) (2004b) Planning policy statement 12, local development frameworks. TSO, London

ODPM (Office of the Deputy Prime Minister) (2005a) Land use change in England: residential development to 2004, LUCS 20. ODPM Pubications, Wetherby

ODPM (Office of the Deputy Prime Minister) (2005b) Land use change in England to 2004: additional tables and figures, LUCS 20A. ODPM Publications, Wetherby

ODPM (Office of the Deputy Prime Minister) (2005c) Planning for housing provision, consultation pape. ODPM Publications, Wetherby

ODPM (Office of the Deputy Prime Minister) (2005d) Local development framework core output indicators, update 1/2005. ODPM Publications, Wetherby

ODPM (Office of the Deputy Prime Minister) (2005e) Core output indicators for regional planning. ODPM Publications, Wetherby

OPDM (Office of the Deputy Prime Minister) (2005f) Local development framework monitoring: a good practice guide. ODPM Publications, Wetherby

Pl&CP Act (2004) Planning and Compulsory Purchase Act 2004, Elizabeth II. TSO, London

SEERA (South East England Regional Assembly) (2003) Regional monitoring report 2003

Syms P, Knight P (2000) Building homes on used land. RICS Business Services Ltd., Coventry

Syms P (2001) Releasing brownfields. Joseph Rowntree Foundation

www.buildingforlife.org/apply/buildingforlife.aspx?bfl=true&contentitemid=384&aspectid=15, 14th Feb. 2006

www.englishpartnerships.co.uk, 10th June, 2004

www.sustainable-development.gov.uk/performance/24.htm, 20th Dec., 2005

Augmented Reality and Immersive Scenarios in Urban Planning

Ingo Wietzel, Hans Hagen, and Gerhard Steinebach

1 Introduction

Urban planning is affected by fundamental changes in society and government. As a discipline and practice it must to cope with new technical and juridical challenges on content, methodical, and procedural levels. Classical decision-making and assessment method must be modified, otherwise they will not be able to handle the future challenges.

Urban planning is not a classical lab discipline. The urban planning laboratory has to be based on augmented reality techniques. But this is not all; urban planning is not a virtual computer game. To really support human creativity we have to create immersive scenarios. This is what this paper is about.

2 Assessment and Decision Making in Urban Planning

2.1 "Classical" Decision-Making and Assessment Methods in Urban Planning

Different (spatial) aspects have to be balanced in urban planning. There are always different claims and ideas in the initial phase of a planning process. There is not one optimal way and method in hardly any case, but very many (slightly) different solutions. Political constraints and normative objectives scale down the breadth of possible solutions quite a bit. What are useful constraints? Differentiated quantitative and qualitative decision-making and assessment methods are indispensable. Comprehensibleness and transparency are the key factors. But there are no mathematical and general models. The job of mathematical models and algorithm

I. Wietzel(✉)
Technische Universität Kaiserslautern, Lehrstuhl Stadtplanung, Pfaffenbergstr.95, 67663 Kaiserslautern, Germany, E-mail: wietzel@rhrk.uni-kl.de

G. Steinebach et al. (eds.), *Visualizing Sustainable Planning*,
© Springer-Verlag Berlin Heidelberg 2009

(in principle, the whole of computer science support) is to support and strengthen creativity and not to replace it. In the past, a comparison was made concerning the suitability of decision-making and assessment methods used in spatial planning and with the help of the formal requirements for useful assessment methods. The suitability classification with regard to the transparency and comprehensibleness of the methods is especially remarkable. This criterion is classified with almost all quantitative methods as doubtful to very doubtful (Scholles 2005). This means, that the responsible actors have to handle with decisive factor positions they cannot understand. So, it is not clear if the suitability of textual works, tables, and atlases as classical representational forms is very useful in the mediation of the results.

2.2   Changing Basic Conditions and Derived Consequences for Decisive Factor Positions

There have been changes in the basic conditions of the spatial planning process over the last couple of years. The reasons are state-related and society-related as well as based on technical and juridical requirements. There are three aspects to the increasing complexity.

- The number of actors with different interests, needs, and values.
- The number of variables to be taken into consideration as well the interrelations among the different aspects of the spatial development process.
- The social future orientation demands a huge number of alternative developing options (Steinebach and Mueller 2006).

The changes of the underlying constraints need to the following facts:

- The available data bases increase constantly.
- The developments in the area of information and communication technologies offer new possibilities.

Because of these circumstances the spatial planning system is forced to accelerate the expiries in planning and decision making. But in Germany, these are still too protracted and too inflexible; as a result, the requirements for the decision-making process increase. Therefore it is important to develop new representation forms and methods that are understandable for all involved actors.

2.3   Some Details of Human Perception

The world is three-dimensional, and everybody is accustomed to operating in this environment in a natural way. The three-dimensional methods improve our orientation in three-dimensional space, our identification of objects and locations. The habits and behavior of a person in three-dimensional space is important – it is a supporting component in the perception process. The basis for our perception of objects and the surrounding space are physical sensations. These are the raw materials of the human experience. Perception is a process of admission, forwarding, selection, interpretation, and association of sense-oriented information. If the person is able to associate the perceived with known mental patterns, or models, he or she is able to react to that intuitively. According to the results an investigation conducted by the U.S. Department of Agriculture and Forest Service, the sense-related shares of perception are as follows: vison 87%, hearing 7%, sense of smell 3.5%, sense of touch 1.5%, sense of taste 1% (Weidenbach 1999). This suggests that visual impressions decisively determine spatial perception. A reduction of the three-dimensionality of the surroundings through verbal descriptions or two-dimensional pictures always causes a reduction of information. This loss of information requires a mental reconstruction. The mental reconstruction uses patterns or model associations. Divergences between the original object and the mentally reconstructed objects are inevitable. In urban planning, the mental reconstruction of objects creates variations among individual descriptions. Greater demands on the mental reconstruction lead to greater divergences of the reconstructed results. This is important for planning decisions, in which a rising number of actors are involved with different knowledge bases.

2.4 Forms and Possibilities of Representation in Urban Planning

Different forms of representation are used in urban planning. This means text works, including tabular pictures as well as plan works or physical models. In addition, digital representations have been used for some years now. As a rule, combinations of different representations are chosen in urban planning to mediate information. The form representation contributes decisively to the comprehensibleness of planning, including the results of assessment methods. The value in the decision-making process is accordingly high, but not every urban planning situation needs a three-dimensional representation for the mediation of information. Certainly, the third dimension is getting more and more important for a "construction-structurally referring consideration." For example, this concerns the building density by areas, figure quality of accommodations and parts of the town, shadow situations of buildings and free surfaces or noise reduction by certain structures. Architectural measures cannot be sketched to the scale 1:1. The abstraction and reduction in different scale steps and visualization technologies are very useful and absolutely necessary. However, a reduction of information may go along with this. In addition, analogous maps or plans have the disadvantages of not allowing 3-D representation. They are fixed in each case to attached cuttings and views, therefore a change in the point of view or the scale is not possible. But the aforementioned disadvantages can be overcome by using digital representation. The use of 3D city models as well as a walkthroughs are more common and are taking an increasingly bigger role in the area of simulation. They can be used as a calculation base for noise grid mappings. Nevertheless the biggest deficiency in the digital representation lies with the viewing-media. Because of the size of the display only partial cuttings can be looked at, either by zooming in or zooming out or by walking along, and this tends to cause a loss of information. At the moment multi-screen solutions, with four or more horizontally and vertically arranged displays, are still quite expensive for workplaces. The grid effects originated in these solutions influence the perception of the general view. The application possibility of projectors is limited by the native picture resolution. Without using special hardware and software, a spatial representation remains actually two-dimensional, which also causes a loss of information. Nevertheless, it is impossible to create a virtual or augmented planning situation without mental reconstruction.

3   Augmented Reality Technology

In 1994, Milgram and Kishino sketched a model in which they defined a continuum from more real to virtual surroundings to illustrate the connection of reality and virtuality. Mixed reality with different overlapping intensities of the real and the virtual world is the outcome of his model. The overlapping of the virtual space with single elements of the real surroundings has to be understood as augmented virtuality. Augmented reality means that the real surroundings are overlaid or complemented by virtual elements. This assumes a digital data shade of the surroundings as well as the real objects to be overlaid. This data shade contains geometrical information and the position of the object in a co-ordinate system. Information, for example, about material, age, and appearance can be applied in a modified condition in this data shade. Single virtual reality objects are faded in by the augmented reality technology as additional elements in the real surroundings. At the moment numerous augmented reality applications exist in medicine, military science, aviation, and astronautics. Included in these applications are research and development, production, manufacturing, and assembly as well as service and maintenance. However, most applications are still in the prototype phase. An augmented reality system is characterized by three features (Azuma 1997)

- combining real and virtual objects in realistic surrounding,
- real time interactivity,
- adjustment and registration of real and virtual objects.

Not only the classical supplement is possible by virtual objects, but real objects from the perceived surroundings can be extracted. So it is possible to cover an existing building and to remove it from the perceived scene.

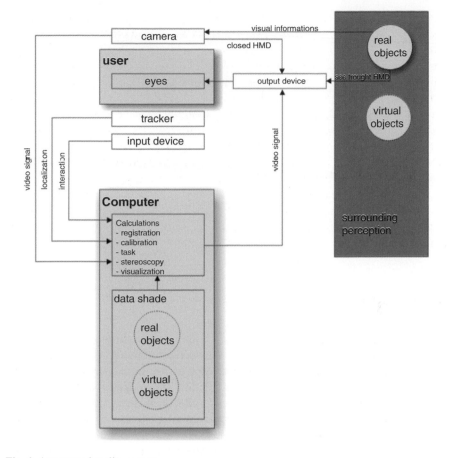

Fig. 1 Augmented reality system

4 Immersive Scenarios

Research and development in the hardware and software sector have vastly increased the possibilities to visualize data and information during the last few years.

The next step shows the development of systems that will enable the user to perceive virtual surroundings so vividly that it will be difficult to distinguish between real and virtual, or augmented, surroundings. This process is called immersion. Virtual objects become apparent components of reality. When the user operates freely and intuitively he or she is in an immersive scenario. The immersion allows a motor and sense perception of the representation, as well as the opportunity for the user to interact in the virtual surroundings. The immersion is

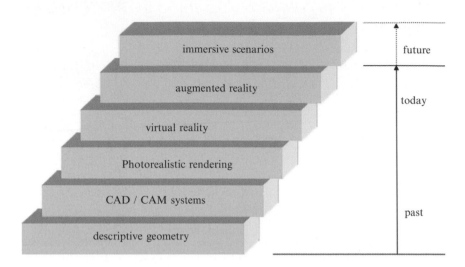

Fig. 2 Development in visualization of data and information (Steinebach et al. 2007)

created by entirely or partially fading out the real surroundings, the number of the appealed sensory irritations, and the liveliness of the virtual surroundings. The degree of the immersion is determined by the media quality of the system. The stimulation of all sensory modalities is the ideal of an immersive-created system, but it is not always necessary to consider all aspects in every case. However, it is important to consider that if sensory stimuli are missing, the degree of immersion decreases. The degree of immersion of a scenario is determined to three essential factors

- the cognitive abilities of the user,
- the level of detail of the scenario, and
- by the interaction possibilities of the user.

The user can have an effect on the scenario as well as the scenario can have an effect on the user. Ideally, the user can move fluidly between reality, augmented reality, and virtual reality. The following requirements are essential in generating an immersive scenario:

- the use of the third dimension
- the stimulation of sensory modalities
- guarantee of the intuitive functionality

Innovative technologies and methods for the intuitive interaction with virtual objects have been in use for some time now, yet it's still a new development in the field of graphic data visualization and the visualization of information.

5 Possible Applications of the Augmented Reality Technology and Immersive Scenarios in Urban Planning

5.1 Basic Separation of the Operational Areas of Augmented Reality and Immersive Scenarios

How can immersive scenarios and augmented reality be used as a rational application in urban planning?

5.1.1 The Application of Augmented Reality Technology and Immersive Scenarios to the Result Visualization

Up to now augmented reality applications have been used for visual representation of relevant information only. Real interaction possibilities in augmented reality applications have existed only in rudimentary form up to now. These techniques are not really necessary in some operational areas, for example, in the pure mediation of information for decision making. The visualization of information in the real or known surroundings is important for the conviction ability. So, relationships can be ascertained from the surroundings directly. The mental reconstruction of spatial relationships is not necessary anymore. As a result, visualizations of augmented reality technology differ substantially from conventional representation forms in their descriptiveness and comprehensibleness. The application potential of the augmented reality technology in the spatial planning process can be considered very high. The planning process is marked by a huge number of decision-making situations. The application of augmented reality technology in the decision-making processes opens up the possibility of arranging the architectural and formative consequences of plans and measures in the real space more realistically. Contrary to classical model forms, a realistic simulation of the architectural and formative plans and measures is possible.

5.1.2 Immersive Situation Representations

The users of immersive scenarios should be able to see and feel the results and consequences of planning. The key criteria in immersive scenarios are the intuitive interactions and the perceptual experience. The user should be able to interact equally with real objects as well as with virtual objects. Changes can be directly recognized. The functionality of the virtual object changes from pure consideration to usability. The construction of virtual objects takes place almost exclusively in classical 2D forms at the moment, and augmented reality shows the results in the real surroundings. But with immersive scenarios, object modeling as well as object-attribution could occur within the scenario intuitively. Urban planning could benefit in many areas within the planning process from such a situational representation. On this occasion, the immersive experience of plans and their consequences is less important than intuitive interaction on site. For example, a digital stock-taking would be stated with the inspection of an investigation space. The objects of the stock-taking could be taken up directly in a digital plan basis by means of intuitive actions, for example, by drag and drop of predefined attributes or linguistic input.

The support of the classical plan-making by immersive scenarios is also possible. Sketching of structures could be performed by intuitive interaction with virtual objects on site. So there comes up the possibility to use the third dimension of the planning space as well as the original scale already in the planning stage. Thereby deficits would be detectable earlier in the process. It would be possible to fade in all aspects relevant in the planning process, as well as to carry out simulations on a real-time basis on site. The results could lead to new planning alternatives and planning variations. So an optimization of the planning alternatives as well as a qualification of decisive factor positions is also possible. In addition, there could originate a new form of cooperative work in which several partners could have an effect on different planning variations directly on site.

Basically, the real potential of augmented reality technology is in planning objects and considering areas

- that show an immediate relation to the aspect of the three-dimensionality of space, and
- that show some potential to influence the actors qualification to cope with the comprehensibleness in certain decisive situations.

Based on this, immersive scenarios are useful,

- if complicated structures cannot be completely illustrated by pure visual representations
- when the planning situation requires a high degree of interaction.

5.2 Classification of the Virtual Supplements in Urban Planning

Which kinds of the content supplement can be basically classified? Human vision has the highest value as already demonstrated. Accordingly the visualization from originally visible and originally not visible information represents the essential differentiation for the classification. So, three main-classes of information supplement can be stated.

5.2.1 The Supplement by Originally Visually Discernible Information

Augmented reality technology and immersive scenarios allow realistic simulation even on a 1:1 scale in real surroundings. This is already a very direct form of simulation, since the future situation is simulated and visualized in an already existing environment. This can be called a digital and adaptively extendable development method.

5.2.2   The Context-Sensitive Reduction of Original Information

It can be useful to fade out information that is not directly connected with the situation we are looking at. Another possibility consists in highlighting certain features. Both ideas can be implemented by augmented reality technology and immersive scenarios. The computer-aided, visual reduction of information can only be carried out technically by fading the real surroundings with virtual objects. Considering many focus aspects in urban planning can lead to a situation that is too complex to be analyzed efficiently. If this is the case, a context-sensitive and user-driven reduction of information or a visual increase of the relevant information has to take place. By using the augmented reality technology or immersive scenarios different subranges of the real spatial surroundings can be emphasized more strikingly. An example of this is the visual emphasis on all building edges in a defined area or the fading of complicated structured building facades.

5.2.3   The Supplement of Visually Indiscernible Information

Object information that is visually indiscernible in the real surroundings can be visualized by using augmented reality. In the simplest case this happens by inserting textual form. In addition, it is possible to visualize elements that are discernible in reality but not visually, such as pollution. A precondition for this is the production of effect models. It is possible to simulate effect structure of the actual situation and the target situation based on these models, to transform the results, again, in visually discernible information. Usually an allocation of the attributes, stamping, parameters, measuring unities, etc., is necessary for symbols or color values. In addition, it is possible to visualize indiscernible urban planning aspects. For example, through the textual insertion of additional building data more complicated structures can be visualized such as catchment areas, airflows, pollution, or noise. The visualization of invisible aspects can be incorporated in actual situation, and in simulations of planning alternatives. In principle, it is also possible to visually represent visually indiscernible information ones. Certainly, this affects the degree of immersion. According to efficiency of the immersion-creating system, the sensory of modalities of the person has to be appealed originally and separately. An example should clarify the difference: classical noise calculations can be carried out by known and standardized sound propagation models to judge the expected noise load on a property affected by traffic control changes. Augmented reality technology is necessary to transform noise into visually discernible information. The simulated noise effect can be illustrated by color scales in the surface or in space. In addition, it is possible to fade in comparative references through symbols. By using an immersive scenario hearing would be appealed to directly. This can happen, for example, by the recording of calibrated authoritative noise charges or by the acoustic simulation of the noise load. Theoretically, the possibilities in using immersive scenarios are various and flexible. Immersive scenarios support creativity.

6   Results and Outlook

The trends described above amplify a need for sophisticated bases for decision making. On this occasion, one challenge will be to provide the amount of data and information necessary in a form that minimizes the divergences of individual perception and subsequent reconstruction. Moreover, augmented reality as well as immersive scenarios can play an essential part, because they illustrate planning situations and planning alternatives in a form that is comprehensible to human perception. Today augmented reality methods are still in the prototype stage, research and development are still necessary, especially in hardware, tracking algorithms, and application software. Several research projects have already been carried out, namely, the @Visor Project of the German Research Center of Artificial Intelligence (DFKI Kaiserslautern). The development of immersive scenarios depends on interdisciplinary research, including the areas of information and data visualization, perception psychology, and HCI (humanly computers interaction). Urban planning specialists must come up with precise guidelines in order for further developments in augmented reality, immersive scenarios, and human computer interaction to be possible. Computer scientists and urban planners have to cooperate from the very beginning, especially in designing the human computer interface.

7   Bibliography

Azuma R T (1997) A survey of augmented reality. Presence: Teleoperators and Virtual Environments

Milgram P, Kishino F (1994) A taxonomy of mixed reality visual displays. IEICE Trans. Information Systems (E77-D):12

Scholles F (2005) Bewertungs- und entscheidungsmethoden. In: Akademie für Raumforschung und Landesplanung. Handwörterbuch der Raumordnung, Hannover

Steinebach G, Hagen H, Allin S, Scheler I (2007) Objective function to generate a planning alternative. Urban Planning and Environment (UPE 7) Symposium in Bangkok

Steinebach G, Müller P (2006) Dynamisierung von planverfahren der stadtplanung durch informations- und kommunikationssysteme. In: Schriften zur Stadtplanung Band 4, Technische Universität Kaiserslautern

Weidenbach M (1999) Geographische informationssysteme und neue digitale medien in der landschaftsplanung. Logos-Verlag, Berlin

II
Environmental Issues

Urban Meteorological Modeling

Joseph A. Zehnder and Susanne Grossman-Clarke

Introduction

The term model, as used in the context of meteorology and climate, refers to a computer code that numerically integrates a set of equations that govern the evolution of the state of the atmosphere. One starts with initial conditions that are ultimately derived from observations but in some cases come, in part, from the output of a larger scale, coarsely resolved model. The initial wind, pressure, temperature, and moisture are specified and the model equations are used to advance these quantities forward, thus providing a representation of the state of the atmosphere at later times. The evolution is determined in part through a mutual interaction between the model variables, but also through external forcing (e.g., solar radiation) and interactions with the earth's surface, which provide fluxes of heat, moisture, and momentum.

The basic equations constituting a meteorological or climate model are the Navier-Stokes equations,

$$\frac{\partial \vec{u}}{\partial t} + \vec{u} \cdot \nabla \vec{u} + 2\vec{\Omega} \times \vec{u} = -\frac{1}{\rho}\nabla p + \vec{g} + \vec{F} \tag{1}$$

which determines the three-dimensional velocity $\vec{u}$, an equation that represents the conservation of mass (continuity)

$$\frac{\partial \rho}{\partial t} + \vec{u} \cdot \nabla \rho = -\rho \left(\nabla \cdot \vec{u} \right) \tag{2}$$

J.A. Zehnder(✉)
Department of Atmospheric Sciences, Creighton University, Hixson-Lied #541,
2500 California Plaza, Omaha, NE 68178-0002, USA, E-mail: zehnder@creighton.edu

G. Steinebach et al. (eds.), *Visualizing Sustainable Planning,*
© Springer-Verlag Berlin Heidelberg 2009

and the conservation of energy as represented by the First Law of Thermodynamics

$$\frac{\partial \Theta}{\partial t} + \vec{u} \cdot \nabla \Theta = H \tag{3}$$

In these expressions, p represents the pressure, ρ the density (mass per unit volume), g the acceleration due to gravity and $\vec{\Omega}$ the earth's rotational axis. In order to provide a closed system of equations we introduce an equation of state, the ideal gas law,

$$p = \rho RT \tag{4}$$

which relates the temperature T, density, and pressure and the potential temperature

$$\Theta = T_0 \left(\frac{p_0}{p} \right)^{R/c_p} \tag{5}$$

appearing in the First Law of Thermodynamics. Forcing of the system is represented by $\vec{F}$, which represents sources and sinks of momentum due to surface friction and turbulent diffusion within the interior of the atmosphere and H, which represents sources of heat such as solar radiation, infrared radiation, conduction from the surface and latent heat due to the condensation of water vapor. There are a myriad of other processes present in the real atmosphere that are, in general, incorporated into the forcing terms F and H and the details of these are beyond the scope of this presentation. Some details relevant to the modeling of the urban atmosphere will be presented in the following sections.

Technically, we are describing motions and processes that occur within a thin, spherical shell (i.e., the atmosphere around earth). This fact can be accounted for through a proper choice of coordinate system and map projections that incorporate the earth's curvature. Further, variations in elevation of the surface are included through the introduction of a terrain following vertical coordinate. One example commonly used is the so-called "sigma coordinate,"

$$\sigma = \frac{p - p_s}{p_t - p_s} \tag{6}$$

where p is the pressure and p_s, p_t are pressures along the surface and top of the atmosphere respectively. The earth's surface has a constant coordinate value of $\sigma = 0$.

Unless fairly restrictive simplifying assumptions are made, the equations must be solved numerically on a high-speed computer. A commonly used technique converts the equations into what is known as finite difference form. A three dimensional lattice of grid points is defined as shown in Fig. 1 and suitably

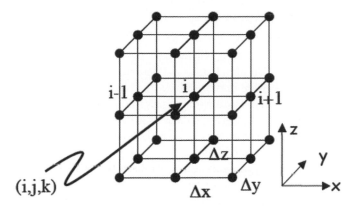

Fig. 1   Schematic representation of a grid point (i) and adjacent points on the three dimensional lattice

stretched in each direction to account for map projections and the terrain following coordinate.

A value of each variable is then assigned to each grid point. The value of a quantity (say u, one of the velocity components) at some time (say t + Δt) is given in terms of the value at an earlier time (t − Δt) and the current value. Horizontal gradients are represented through finite differences at adjacent grid points. This can be more clearly visualized through a finite difference version of a single component of the Navier-Stokes Equations

$$
u_{i,j,k}^{t+\Delta t} = u_{i,j,k}^{t-1} + 2\Delta t \left[
\begin{array}{l}
u_{i,j,k}\dfrac{u_{i+1,j,k}-u_{i-1,j,k}}{2\Delta x} + v_{i,j,k}\dfrac{u_{i,j+1,k}-u_{i,j-1,k}}{2\Delta y} + \\[2ex]
w_{i,j,k}\dfrac{u_{i,j,k+1}-u_{i,j,k-1}}{2\Delta z} - \dfrac{1}{\rho_{i,j,k}}\dfrac{P_{i+1,j,k}-P_{i-1,j,k}}{2\Delta x} - \\[2ex]
fv_{ijk} + F_{x(i,j,k)}
\end{array}
\right] \tag{7}
$$

Here, u, v and w represent the three components of velocity and other terms are as described above. Similar forms of the other equations allow other variables to be advanced forward in time, and a sequential replacement of old equations with updated values allows the model state to be advanced forward in time. There are numerous details, such as limitations on the time increment (Δt) imposed by the spacing of the lattice and stability properties of the integration scheme that are beyond the scope of this presentation. The interested reader may consult a text on numerical weather prediction (e.g., Haltiner and Williams, *Numerical prediction and dynamic meteorology*, New York: Wiley and Sons, 1980).

Prediction Versus Downscaling

The discussion above relates to using meteorology/climate models in a predictive sense. That is, providing an initial state and determining a final state based on the numerical integration forward in time. While this is a very common utilization, one generally runs the model on a grid with elements that are more closely spaced than the observations (which are usually taken from the operational weather service network). This mismatch in spacing can be addressed by interpolating the observations onto the model grid using an objective analysis scheme. These schemes tend to blend the observations smoothly and continuously onto the model grid. This same technique, the mathematical interpolation of observations, is also used in data analysis and visualization. However, the true character of features at scales below that of the observations is not likely to be captured by this method.

An option that provides a better representation of fields on scales below that of the observations allows the model physics to provide an additional level of detail. One initializes a model with the coarse observations and allows details to be provided through the model physics and accompanying conservation laws. In this case, we are using the observations as a "first guess" and allowing the evolution as governed by the model physics and forcing to provide physically and dynamically consistent values of the variables on the model grid scale.

In general, numerical weather forecast models such as those run operationally in support of weather services worldwide are run on 1–10 km grids, while observations are typically on the 10–100 km scale over land, and much greater over the oceans. Long-term climate prediction models are run on 10–100 km grids, and, while this is comparable to the spacing of the observing networks, details at finer scales are desirable in many applications.

With weather prediction models, one often performs a "cold start," in which observations are interpolated onto the model grid and improved values are generated on the grid scale over a transient period that is usually some fraction of a diurnal cycle. The start-up time varies with grid spacing and is inversely proportional to the grid size (i.e., shorter times for higher resolution models). With climate simulations, one uses the output of a climate model as input into a fine resolution weather prediction model. In this way, details of the long-term climate may be determined on a spatial scale that cannot be provided by the weather prediction model alone. In either application, an accurate representation of the physical processes is necessary.

Parameterization Of Subgrid Scale Processes

Variability and scales of motion in the atmosphere occur on scales from the planetary to the molecular. Regardless of the grid scale one chooses to use, there are processes occurring on the scales smaller than can be resolved. In order to include these processes into the model, the cumulative effect on the resolved scale must be represented. For example, the dissipative effect of turbulent motions on the resolved scale velocity (say u) may be represented as

$$F_{rx} = K\left(\frac{\partial^2}{\partial x^2} + \frac{\partial^2}{\partial y^2} + \frac{\partial^2}{\partial z^2}\right)u \tag{8}$$

where K is an empirically-derived effective viscosity. Here, the model does not resolve the individual turbulent motions, but rather the net effect on the model scale flow is included. This formulation is valid as long as the model grid spacing is larger than the turbulent scale (i.e., >1 km).

A similar situation occurs when including the effects of clouds in models. A typical cumulus cell is on the order of a 1–10 km in width. In climate model simulations, the cumulative heating and moistening effect of the clouds on the scale of the model are important, while the individual up- and down-drafts would cancel each other out. For weather prediction models, and particularly when used at the 1 km range, the influence of individual cells and the momentum transfer as well as the heating and moistening must be included. It is not possible to create a general-purpose cloud scheme that is applicable across all scales.

Urban Specific Processes

A goal of meteorological modeling in urban regions is to provide an accurate representation of the urban environment on scales from the regional, to the neighborhood, to the microclimate (building level) scale. Owing to the complexities of representing subgrid scale processes, it is not possible to design a single purpose computer code that covers all these scales. Also, the highly heterogeneous nature of urban regions introduces an additional level of complexity when compared to modeling the surface/atmosphere interaction over the oceans or relatively homogenenous land surfaces such as forests or deserts.

The energy exchange between the surface and atmosphere occurs in part through a transfer of momentum that is due to turbulence induced by surface friction and/ or by random buoyant motions. The turbulence also transports heat and water vapor. Incoming solar radiation is partially reflected back to space and partially absorbed. The fraction of the incoming light reflected is represented by the albedo of the surface. The radiation that is absorbed heats the surface. The resulting temperature change depends on the surface's heat capacity, which relates the heat absorbed to the temperature change of the material and the conductivity, which determines how effectively the heat is transmitted through the material.

The surface also emits energy according to the Stefan-Boltzmann law, which relates the emitted energy to the fourth power of the surface temperature through a material-specific constant of proportionality, the infrared emissivity. Built materials tend to be high in density and efficient absorbers and conductors of heat, so solar radiation is absorbed and stored and then remitted at night.

Another factor that differentiates urban from natural surfaces is the availability of water. Urban areas are composed largely of impervious surfaces, hence rainfall tends to run off and be transported away rather than being absorbed into the ground where it falls. As a result, urban areas tend to have less available moisture than adjacent natural areas[1].

Surface energy schemes are a component of meteorological models that calculate the energy exchange between the earth's surface and the atmosphere. Commonly, a single value of the surface parameters (albedo, heat conductivity and capacity, emissivity, roughness) is used to characterize a model grid cell, with that value being an average or dominant value over the grid cell. Some differences related to the scale of the grid can be seen by examination of Fig. 2, which shows a visible satellite image of parts of Phoenix as they are captured by two different resolutions of the model grid. Say, for example, that we wish to characterize the surface roughness over the regions in Figs. 2a and 2b. The four partitions will be smoother and more have similar values in Fig. 2a than in 2b. Also, the available water in the upper right quadrant in 2a will be higher than in the other sections. Fig. 2b shows a portion of the domain in 2a and would represent the surface in a higher resolution model. Here, there is much more heterogeneity between the quadrants. Also, owing to the larger proportion of highly reflective rooftops in the lower right quadrant, this portion of the model domain would have a higher effective albedo. Fig. 2 serves to illustrate some of the challenges in characterizing the land surface properties in urban areas, particularly when high resolution grids are used.

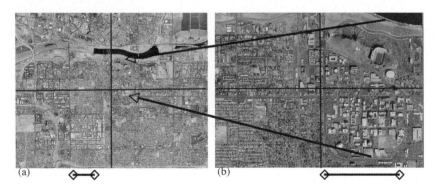

Fig. 2 Visible satellite image of the Phoenix metropolitan area at varying scales. The bar under each frame is 1 km in length

---

[1]This is not true in arid and semi arid regions due to irrigation. The implications of this will be discussed later.

Community Models and Customization for Specific Uses

Owing to the complexity of meteorological and climate models, it is not wieldy for users to develop individual codes for each specific application. Instead, collaborative efforts have resulted in so-called "community" models that have been developed over time with contributions from many users. One of these models is referred to as the Penn State/National Center for Atmospheric Research Fifth Generation Mesoscale Model, or MM5 for short. MM5 contains schemes to represent the relevant physical processes and multiple options to account for the scale dependence of the parameterizations, as discussed above. While the MM5 model is quite complete in terms of processes related to natural land and water surfaces, there are physical processes related to urban areas that are not represented in the community version of the model. The representation of land cover types in the default version of the model is also deficient. Part of the deficiency in the land cover representation, particularly in terms of urban regions, stems from the model grid sizes typically used.

The land cover data set used in MM5 is based on a 24-category scheme developed by the United States Geological Survey (USGS). This classification scheme includes a variety of vegetation types, (forests, grassland and agriculture) water and snow/ice, and urban/built up land. The data is available on a roughly 1 km grid, and MM5 takes the specified model grid, determines a dominant land cover type for that grid and classifies the entire grid cell as that type

The overall features in the dominant land cover are consistent as one increases the resolution, but there are some peculiarities involving the urban land cover. At the coarsest resolution, a single grid cell represents the Los Angeles metropolitan area. None of the other urban areas in the southwestern United States occupies a large enough area to dominate an 81 km grid cell. As the model resolution increases, one eventually sees the Phoenix metropolitan area, which appears as a single grid cell at 27 km. It is only at the highest model resolutions that the features of the area begin to appear.

Further details of the urban are shown in Fig. 3, which was derived from high resolution LANDSAT images (Stefanov et al. 2000). The satellite imagery used to develop this land cover map is based on the Landsat Thematic Mapper, which is available on a 30 meter grid, hence there is more detail available than in the USGS land cover. The urban area is characterized by three land cover types, mesic (moist) and xeric (dry) residential and commercial/industrial. Heterogeniety of the surface throughout the metropolitan area is clearly seen. Also, areas to the west and south of the metropolitan area that are considered agricultural are only partly active, with other areas lying fallow. These are typically areas that are slated for residential development.

A comparison of the actual land cover with that in MM5 is provided by Grossman-Clarke et al. (2005). There are large portions of residential areas in the Phoenix metropolitan area that are missing from the USGS classification scheme. Areas that are classified as either natural vegetation or agriculture in MM5 are in

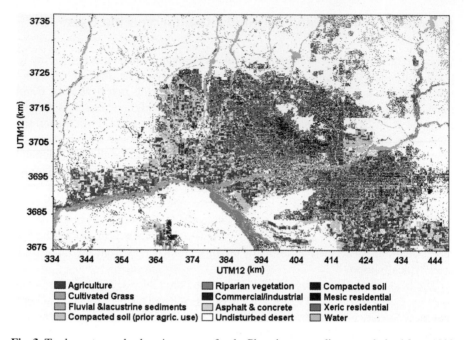

Fig. 3 Twelve-category land use/cover map for the Phoenix metropolitan area derived from 1998 Landsat Thematic Mapper reflectance data and ancillary datasets (after Stefanov et al. 2001). The spatial resolution is 30 m/pixel

fact residential (i.e., in the southeast portion of the region). Also the USGS scheme classifies the areas surrounding the metropolitan area as shrubland when in fact it is dry and sparsely vegetated desert. This misclassification can have important ramifications since MM5 assigns physical characteristics to the grid cells via a look-up table once the land cover type has been assigned. Some example values are provided in Table 1.

The urban land use classification in the USGS data set is based on a US Department of Defense data set and is correct circa 1966. In a sense, this coarse and out of date classification was unimportant until recently when the models could be run on sufficiently fine scale grids in order to resolve the details of the urban area. Further, the USGS scheme has a single category to characterize urban areas. This limits the ability to represent differences in moisture, thermal properties and roughness that occur in real urban areas. In order to accurately represent the urban environment in the models, it is necessary to address these deficiencies and customize the community version of the model.

A surface energy scheme that addresses some of the deficiencies in the community version of MM5 has been developed at Arizona State University with details of the model formulation given in Grossman-Clarke et al. (2005). This scheme consists of two parts, one being the incorporation of a revised land cover

Table 1 Values of physical parameters for land cover types used in MM5. Values are assigned allowing for seasonal variation of parameter values

Vegetation description	Albedo (%)	Moisture (%)	Emissivity (%)	Roughness length (cm)	Thermal inertia (cal cm^{-2}K^{-1}s$^{-1/2}$)
Urban	18	10	88	50	0.03
Dry Crop	17	30	92	15	0.04
Irrig. Crop	18	50	92	15	0.04
Mixed Crop	18	25	92	15	0.04
Crop/Grass	18	25	92	14	0.04
Crop/Wood	16	35	93	20	0.04
Grassland	19	15	92	.12	0.03
Shrubland	22	10	88	10	0.03
Grass/Shrub	20	15	90	11	0.03
Savanna	20	15	92	15	0.03
Decid. Broadleaf	16	30	93	50	0.04
Decid Needleaf	14	30	94	50	0.04
Evergr. Broadleaf	12	50	95	50	0.05
Evergr. Needleaf	12	30	95	50	0.04
Mixed Forest	13	30	94	50	0.04
Water	8	100	98	.01	0.06
Wetland	14	60	95	20	0.06
Wooden Tundra	14	35	95	40	0.05
Sparse Veg.	25	2	85	10	0.02
Herb Tundra	15	50	92	10	0.05
Wooden Tundra	15	50	92	15	0.05
Mixed Tundra	15	50	92	15	0.05
Barren Tundra	25	2	85	.10	0.02
Snow/Ice	55	95	95	5	0.05

classification into the model and correcting or reclassifying the surrounding natural areas when necessary. The second part involves including additional physical processes into the model to better represent the urban energy balance.

There are three factors related to the urban energy balance that must be included into the model. The first involves the way in which infrared (heat) radiation behaves in urban areas. In general, the sun heats the surface and the surface cools at night by emitting infrared radiation. The heat is emitted in a direction perpendicular to the surface, and in natural areas this is up and into space. In areas with built surfaces there are vertical walls and the radiation is horizontal rather than vertical. The radiation is "trapped" in part by the fact that walls radiate against each other, and this reduces the effective cooling rate. This process is illustrated schematically in Fig. 4.

The effect of radiation trapping may be represented in a quantitative fashion by introducing a sky view factor, Ψ_{sky}. This is a dimensionless number with a value between 0 and 1 that represents the fraction of visible sky at a point in comparison

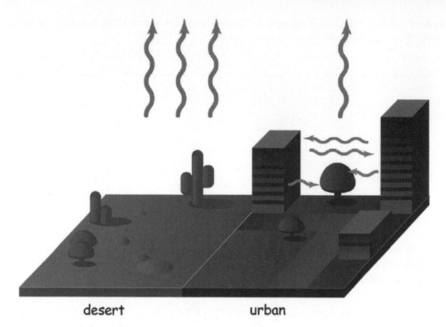

desert   urban

Fig. 4 Schematic representation of radiation trapping by built materials

to that over a horizontal surface. The value may be represented in terms of representative building heights (h) and road width (w) as

$$\psi_{sky} = \left[\left(h / w \right)^2 + 1 \right]^{0.5} - h / w \qquad (9)$$

Representative values of this quantity for the Phoenix metropolitan area are typically in the .75 to .85 range. ψ_{sky} is in essence a scale factor that reduces the outgoing longwave radiation, written as

$$R_{long} = \psi_{sky} \varepsilon_g \left(L \downarrow - \sigma \cdot T_g^4 \right) \qquad (10)$$

where $L\downarrow$ (W m^{-2}) is the incoming long wave (infrared) radiation from the sky, ε_g is the emissivity of the surface, σ (W m^{-2} K^{-4}) the Stefan-Boltzmann constant and T_g is the surface temperature. The reduction in the outgoing radiation results in higher temperatures in the urban area, particularly at night.

Another factor that must be incorporated into the model is the effect of anthropogenic heat. In general, there is heat released from traffic combustion, industry, and residential areas. In some cases the heat release is active (such as discharges from smokestacks) or passive (such as heat leaked from the interiors of buildings to the outside as part of ventilation or poor insulation). In arid and tropical regions the heat transfer from buildings is active via air conditioners. The surfaces of buildings are heated via solar and infrared radiation and that heat is conducted into the buildings. In order to maintain habitable temperatures in the interior of the building heat pumps transfer the heat to the outside (with the addition of a small amount due to the thermodynamic inefficiency of the pump). While the details of these processes are considered sub-grid scale in the context of meteorological models, they can be parameterized simply by assuming a diurnal variation and assigning an amplitude of the heating based on power and energy consumption (Grossman-Clarke et al 2005).

A third factor that must be accounted for is the capacity for heat to be stored by the built material. Concrete is more densely packed than soil, in general, and hence can conduct heat more easily. This heat is conducted back to the surface at night as the surface cools, and provides an additional effective source of heat. This is incorporated into the model through an appropriate choice of coefficients that govern heat transfer between layers in the model below the surface.

The modified version of the MM5 surface energy scheme that will be discussed here is formulated as above. The heterogeneity of the urban surface is represented by introducing two additional urban categories into the model (urban/commercial, dry (xeric) and moist (mesic) residential). The physical characteristics of these categories are adjusted appropriately and the interested reader is referred to Grossman-Clarke et al. (2005) for details. A representation of the land cover as aggregated to a 2 km grid is given in Fig. 5. This should be compared with Fig. 3.

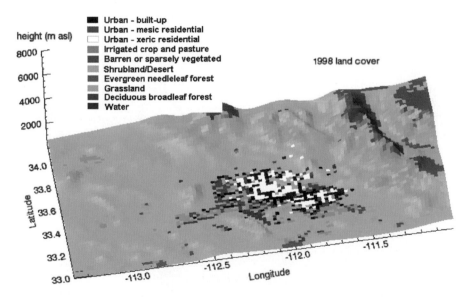

Fig. 5 Land cover and terrain (in meters) from MM5 preprocessor on a 2 km × 2 km grid. Land cover is based on Landsat Thematic Mapper 1998 image derived land use/cover data and three urban land use/cover classes: urban built-up; urban mesic residential (irrigated); and urban xeric residential (arid, partly irrigated). Other land use/cover categories were the same as in the standard MM5 24-category USGS data set

An example of the model performance with the standard version of MM5 and the customized urban version is shown in Fig. 6. This simulation covers a three-day period in June 1998. This time period was chosen since it corresponds to our land cover data set. Also, during the early part of the summer there are typically no weather systems in the region and the temperature variation is governed by local surface heating.

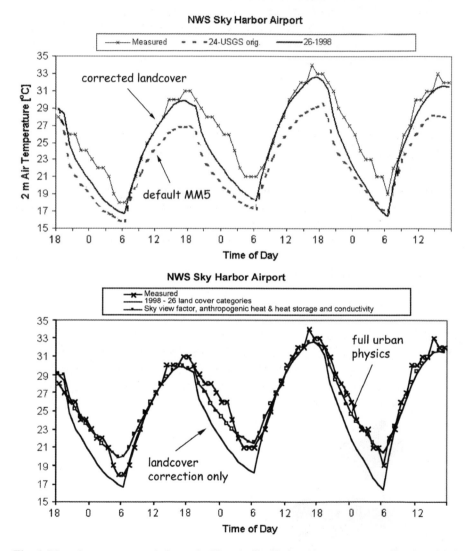

Fig. 6 Diurnal temperature variation at the Phoenix Sky Harbor Airport as measured and modeled for two days: (**a**) observed (-Δ-), simulated using default MM5 (---) and simulated using MM5 with corrected land cover only (–); and (**b**) observed (-Δ-), simulated using MM5 with corrected land cover only (–) and simulated using MM5 with corrected land cover, radiation trapping, anthropogenic heating and heat storage (-o-)

The measured temperatures in Fig. 6 show a clear diurnal temperature variation with a peak at about 6:00 pm local time. The temperature range is about 12 °C (~22 °F) over the course of a day. The default version of the MM5 performs poorly, with temperatures being lower during all points of the day (with the exception of the pre-dawn period). Part of the poor performance is due to improper land cover classification and assignment of parameters. The default version of MM5 has too much available moisture, resulting in unrealistically low temperatures.

Correcting the land cover allows the model to capture the daytime portion of the temperature correctly, but the model still cools too rapidly at night. By including the additional urban physical processes, which primarily provide additional sources of surface heat at night, the full diurnal cycle of temperature is captured by the model.

Aspects of the Urban Heat Island in the Phoenix Metropolitan Area (Past, Present and Future)

A series of model simulations were performed using the urban surface energy scheme described above. Since we must include custom versions of the land cover for current, real time simulations, we can include alternative representations of the land cover and model the evolution of the urban heat island (UHI) with time in the region. Some aspects of the regional character of the UHI are shown in Fig. 7 (maximum and minimum regional 2 m air temperatures).

There are features apparent in Fig. 7 that are contrary to our current understanding of the UHI. In particular, the maximum temperature in the urban area is similar to that occurring in the outlying natural desert areas (lower left corner of each frame) that are at the same elevation. This is in spite of the generally high heat absorption and retention of built materials. The reason for this is an accidental mitigation of the UHI by the presence of surface water (pools, irrigation, artificial lakes and ponds) in the metropolitan area. A substantial fraction of the incoming solar radiation is used to evaporate water, and consequently does not heat the surface.

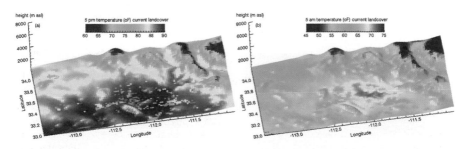

Fig. 7 Surface temperature at 5 pm (maximum) and 5 am (minimum) for June 8 1998

This is in contrast to cities in temperate climates that are generally drier than the surrounding rural areas. The UHI in desert cities is essentially a nighttime effect.

The nighttime character of the UHI is also apparent from Fig. 7. The minimum temperature in the urban core is 10–12 °F higher than in the surrounding natural desert area. The reasons for the elevated temperatures are the combined effects of radiation trapping, anthropogenic heating and heat storage by the built materials. The magnitude of the temperature increase has been documented in a number of studies (Grossman-Clarke et al. 1995 and references therein).

Observational studies also indicate that the UHI has been increasing in the Phoenix metropolitan area, both in terms of the spatial extent and the amplitude defined as the temperature elevation over the surrounding rural areas (Brazel et al 2000). The extent to which future development of the region will alter the distribution and amplitude of the heat island is of interest to state and local officials.

Model predictions of the minimum temperatures for land cover corresponding to 1973, along with a projected land cover are discussed. Fig. 8 shows the land cover distributions for these two cases. Comparing Fig. 8 and Fig. 5 shows an expansion of the urban core since 1973 (Fig. 8a). Large amounts of agricultural land have been replaced by residential development and this trend is continuing in the region. The projected land cover map (Fig. 8b) was provided by the Maricopa Association of Governments (MAG) and is based on the assumption that all available land (i.e., not protected recreational areas or Indian reservations) will eventually be developed. Projections based on current growth rates suggest that this level of development may be achieved by 2040. Surface characteristics were assigned to the projected land cover by assuming that residential density remains the same, but that all new development is of the xeric type. This is consistent with current trends in new construction, but it is not certain if this will continue into the future.

The minimum temperatures corresponding to these land cover distributions are given in Fig. 9. Here we are using the same initial conditions as in Fig. 7, hence the amplitude of the UHI can be compared. We haven't made any attempt to include the influence of observed or projected global warming, but assume that it occurs uniformly across the region.

The changes to the predicted minimum temperature distributions, past and future, are apparent from Fig. 9. The expansion of the extent of the region of elevated temperatures shown in Fig. 9a, and in comparison with Fig. 7b is consistent

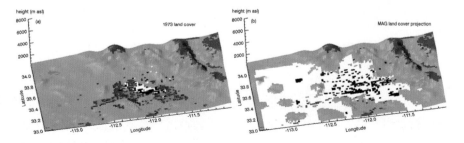

Fig. 8 Land cover for 1973 (**a**) and projected land cover (**b**). See Fig. 5 for legend

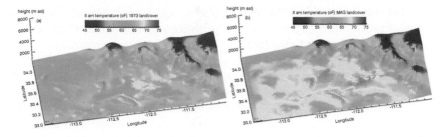

Fig. 9 Predicted minimum temperatures for (**a**) 1973 land cover and (**b**) the Maricopa Association of Governments projection

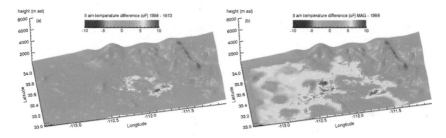

Fig. 10 Differences in predicted minimum temperatures 1998 – 1973 (**a**) and MAG – 1998 (**b**)

with observational studies. The extent of the UHI with the projected land cover is as expected, with relatively uniform and elevated temperatures being present throughout the region.

A more detailed picture of changes in the UHI is obtained by considering difference fields of the temperature, as shown in Fig. 10. The evolution of the temperature between 1973 and 1998 is consistent with observations. That is, there is an increase of between 2 and 6 °F in the urban core, with the largest increases occurring where active agriculture is replaced by residential development. The fact that we can correctly reproduce the observed temperature trend gives us confidence in the prediction for projected land cover. It is interesting to note that the temperatures in the existing urban core are higher by about 2–4 °F in 1998 compared to 1973. That is, in addition to an increase in the area covered by the elevated temperatures, the amplitude of the UHI in the urban core is increasing with time. If this trend were to continue, it could have serious health and environmental implications.

The character of the UHI for the MAG projection is consistent with the previous trend in many respects. That is, there is an increase of between 5 and 10 °F in the residential areas, with the largest increases in areas where active agriculture has been replaced. What is important to note is that there is a negligible increase in the amplitude of the UHI within the urban core under continued development. This result is consistent with observations (Oke 1973) and is due to the influence of an urban/residential "patch" having a finite radius of influence upon the surrounding areas. This effect is particularly pronounced in an area such as Phoenix, where the winds tend to be weak at night. There is little influence of the wind on temperature (i.e., advection) and the temperature trend is governed by local processes.

This chapter outlines some potential uses of meteorological models by planners. In many cases, the models are used in a "real time" forecast mode. That is, observed conditions are input and the state of the atmosphere at some later time is used as guidance for weather forecasts. In planning applications, the models can be used to test future UHI scenarios. That is, the surface characteristics associated with large scale alterations in design (spacing or height of buildings) physical characteristics (albedo, emissivity, heat capacity) or irrigation and other forms of artificial surface water can be input into the model and the resulting response of the temperature (and circulation) can be determined. These changes can then be used to assess power requirements, human comfort and health, and other aspects of the urban environment.

References

Brazel A, Selover N, Vose R, Heisler G (2000) The tale of two climates: Baltimore and Phoenix urban LTER sites. *Climate Res.* 15:123–135

Grossman-Clarke S, Zehnder J A, Stefanov W L, Liu Y, Zoldak M A (2005) Urban modifications in a mesoscale meteorological model and the effects on surface energetics in an arid metropolitan region. *J. Appl Meteor* 44:1281–1297

Oke T R (1973) City size and the urban heat island. *Atmos Env* 7:769–779

Stefanov W L, Ramsey M S, and Christensen P R (2001) Monitoring urban land cover change: an expert system approach to land cover classification of semiarid to arid urban centers. *Remote Sens Environ* 77:173–185

Urban Drainage Modeling and Flood Risk Management

Theo G. Schmitt and Martin Thomas

Abstract The European research project in the EUREKA framework, RisUrSim (Σ!2255) has been worked out by a project consortium including industrial mathematics and water engineering research institutes, municipal drainage works as well as an insurance company. The overall objective has been the development of a simulation to allow flood risk analysis and cost-effective management for urban drainage systems. In view of the regulatory background of European Standard EN 752, the phenomenon of urban flooding caused by surcharged sewer systems in urban drainage systems is analyzed, leading to the necessity of dual drainage modeling. A detailed dual drainage simulation model is described based upon hydraulic flow routing procedures for surface flow and pipe flow. Special consideration is given to the interaction between surface and sewer flow in order to most accurately compute water levels above ground as a basis for further assessment of possible damage costs. The model application is presented for small case study in terms of data needs, model verification, and first simulation results.

Keywords: urban drainage, flooding, dual drainage modeling, hydraulic surface flow simulation, dynamic sewer flow routing.

Introduction

Prevention of flooding in urban areas caused by inadequate sewer systems has become an important issue. With increased property values of buildings and other structures, potential damage from prolonged flooding can easily extend into the millions of dollars. Residents pay service fees and, thus, expect their urban drainage systems to operate effectively without fear of failure due to weather conditions.

T.G. Schmitt(✉)
Technische Universität Kaiserslautern, FG Siedlungswasserwirtschaft, Postfach 3049,
67653 Kaiserslautern, Germany, E-mail: tschmitt@rhrk.uni-kl.de, mthomas@rhrk.uni-kl.de

G. Steinebach et al. (eds.), *Visualizing Sustainable Planning*,
© Springer-Verlag Berlin Heidelberg 2009

However, drainage systems designed to cope with the most extreme storm conditions would be too expensive to build and operate. In establishing tolerable flood frequencies, the safety of the residents and the protection of their valuables must be in balance with the technical and economic restrictions.

According to European Standard EN 752, approved by the European Committee for Standardization (CEN) in 1996, urban drainage systems should be designed to withstand periods of flooding in the range of 10 to 50 years, depending on the type of urban area and traffic infrastructure served.

Flooding in Urban Drainage Systems

Hydraulic Surcharge and Flooding

EN 752 links drainage system hydraulic performance requirements directly to the frequency of flooding. This demands a clear definition of flooding and a distinction from the state – or different stages – of surcharge. According to EN 752 flooding describes a "condition where wastewater and/or surface water escapes from or cannot enter a drain or sewer system and either remains on the surface or enters buildings".

Distinct from flooding, the term *surcharge* is defined as a "condition in which wastewater and/or surface water is held under pressure within a gravity drain or sewer system, but does not escape to the surface to cause flooding." Extended surcharge conditions may eventually lead to a rise in the water level above the surface where water either escapes from the sewer system or prevents surface water from entering the sewer system. Fig. 1 describes different stages of surcharge. Fig. 2 illustrates the phenomenon of surface flooding.

Analysis of Flooding Phenomena

Flooding in urban drainage systems as defined above may occur at different stages of hydraulic surcharge depending on the drainage system (separate or combined sewers), general drainage design characteristics, as well as specific local constraints.

When private sewage drains are directly connected to the public sewer system without backwater valves, the possible effects of hydraulic surcharge depend on the levels of the lowest sewage inlet inside the house (basement), the sewer line, and the water level during surcharge, respectively. Whenever the water level in the public sewer exceeds the level of gravity inlets in the house below street level, flooding inside the house will occur due to backwater effects. In such a case flooding is possible without experiencing surface flooding. In the same way, hydraulic surcharge in the sewer system might produce flooding on private properties, via storm drains, when their inlet level is below the water level of the surcharged storm or combined

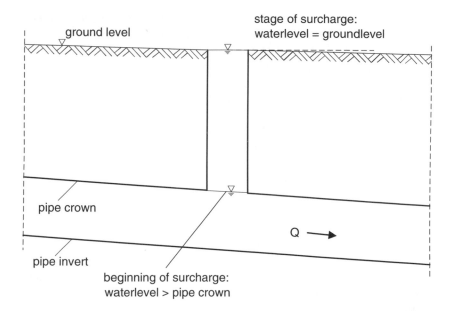

Fig. 1 Stages of sewer surcharge

Fig. 2 Extended street flooding

sewer. In both cases, the occurrence of flooding, being linked directly to the level of inlets versus water level (pressure height) in the sewer, can be easily predicted by hydrodynamic sewer flow simulations, assuming the availability of physical data of the private drains and the public sewer system.

Distinct from the situations described above, the occurrence and possible effects of *surface* flooding depend much more on local constraints and surface characteristics, e.g., street gradient, sidewalks and curb height. These characteristics, however, are much more difficult to describe physically, and these data are usually not available in practice. In addition, today's simulation models are not fully adequate to simulate the relevant hydraulic phenomena associated with surface flooding and surface flow along distinct flow paths.

Due to these deficiencies, the German standard DWA-A 118 "Hydraulic design and simulation of drainage systems" considers flood frequency to be inappropriate for direct computational assessment (DWA 2006, Schmitt 2001). The "surcharge frequency" is established as an additional criterion of hydraulic performance defined as the rise of (maximum) water level at manholes up to ground level. This borderline case of surcharge – the transition from pressurized pipe flow to surface flooding – can be accurately described by current dynamic sewer flow simulation models (Schmitt 2001).

Consequences of Flooding

Flooding in urban areas due to the failure of drainage systems causes extensive damage to buildings and other public and private infrastructure. In addition, street flooding can limit or completely hinder the functioning of traffic systems and has indirect consequences such as loss of business and opportunity. The expected total damage – direct and indirect monetary costs as well as possible social consequences – is related to the physical properties of the flood, i.e., the water level rising above ground level, the extend of flooding in terms of water volume escaping from or not being able to enter the drainage system, and the duration of flooding. With sloped surfaces even the flow velocity on the surface might have an impact on potential flood damage.

Modeling Urban Flooding

In regard to the distinct stages and processes of surcharged sewer system and urban flooding as described above, simulation models for flood risk analysis are required to accurately describe the hydraulic phenomena of surcharged and flooded sewer systems, particularly

- the transition from free surface flow to pressure flow in the sewer pipes,
- the rise of water level above ground level with water escaping from the sewer system,

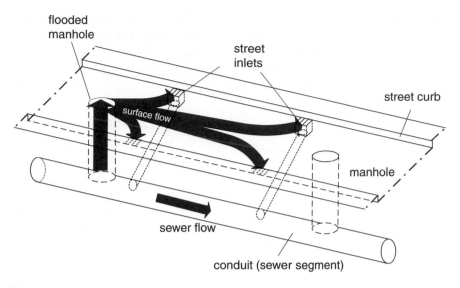

Fig. 3 Interaction of surface and sewer flow ("dual drainage concept")

- the occurrence of surface flow during surface flooding, and
- the interaction between surface flow and pressurized sewer flow.

The consideration of distinct surface flow and its interaction with sewer flow in surcharged sewer systems is denoted as "dual drainage modeling" with flow components on the surface and underground, first described by Djordjevic et al. (1999) and illustrated in Fig. 3.

System Analysis in View of Dual Drainage Modeling

In regard to dual drainage modeling, urban drainage systems are comprised of

- single drainage areas (roofs, streets, parking lots, yards, etc.) where rainfall is transformed into effective runoff depending on surface characteristics (slope, roughness, vegetation, paved/unpaved surface area, etc.);
- distinct surface drainage components, e.g., street gutters, which lead surface runoff to the underground sewer system via inlets,
- surface areas where surface flow might occur in case of surface flooding (e.g., street surface)
- closed underground sewers forming the sewer network (including manholes, control structures, and outlets).

The single areas are connected to the sewer systems via gutters and/or inlets followed by closed pipes. In simulation models the single areas are mostly comprised to sub-catchments that are linked to distinct input elements of the sewer network, generally to the manholes being represented as system nodes. In general, the distinct surface drainage components are not represented by runoff models. The sum of all sub-catchments form the overall catchment area.

For dual drainage simulation single areas need to be further distinguished as follows:

a. in regard to their connection to the sewer system as
    – areas linked completely via closed drains (e.g., roofs)
    – areas linked via surface inlets and closed drains (e.g., parking lots)
    – areas on private sites draining to the street or side-walk surface
    – areas not connected to the sewer system (not runoff-relevant)

b. in regard to possible surface flow as
    – areas not subject to flooding (no interaction between surface and sewer flow, e.g., all roof areas)
    – areas where surface flow occurs and is simulated during flooding
    – areas not to be considered in surface flow simulation

Simulation Model RisUrSim

The dual drainage model RisUrSim has been developed in order to meet the requirements of simulating urban flooding, focusing on the occurrence of distinct surface flow and its possible interaction with the surcharged sewer system.

General Model Features

The overall structure of the model and the general simulation scheme are illustrated in Fig. 4. Following the transformation of rainfall to effective runoff by interception and depression storage simulation, surface runoff is handled in two parallel simulation modules. The module RisoReff is applied to areas where surface flooding as well as interaction between surface flow and sewer flow can be excluded (e.g., roofs, closed private ground). Here, surface runoff enters the sewer system at defined inlets (uni-directional). The module RisoSurf computes surface flow based upon a simplified representation of the shallow-water equations using GIS-based surface data, e.g., street area and slope (longitudinal and lateral), gutters, culvert height, buildings and other features relevant for surface flow patterns. Dynamic sewer flow routing is applied for all underground drainage elements in the module HamokaRis. The model allows *bi-directional exchange* of flow volume between surface flow module RisoSurf and sewer flow module HamokaRis at defined exchange nodes. The bi-directional exchange is realized by interpreting inlets to the sewer system as possible sinks or sources in the mathematical model of both, surface and sewer flow simulation.

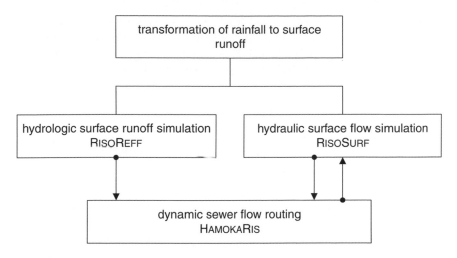

Fig. 4 Interaction of surface and sewer flow ("dual drainage concept")

Rainfall-Runoff Simulation

The RisUrSim model first transforms rainfall into effective runoff using standard methods for interception, depression storage, and soil infiltration (pervious areas only) as described in the literature (e.g., Akan and Houghtalen 1993; Ashley 1999). Surface runoff would then be handled in distinct detail depending on the specific situation of a single runoff area.

For areas not considered for detailed surface flow simulation, e.g., roof areas, RisoReff uses a unit-hydrograph method to compute surface runoff as input to the sub-surface sewer system ("uni-directional flow").

Hydraulic Surface Flow Modeling

The RisoSurf approach includes detailed hydraulic considerations for areas where surface flow occurs. Hydraulic (surface) flow modeling is generally based upon conservation laws of fluid flow expressed in the Navier-Stokes equations. The fact that in surface flow the vertical dimension is much smaller than typical, horizontal scale allows a simplified two-dimensional representation – the "shallow water flow equations" (Hilden 2003).

The application of this detailed hydraulic method would be restricted to small areas only. Therefore, it only served as a benchmark for a further simplified 2-dimensional approach where Manning's equation is used as an empirical flow formula. Neglecting the terms of inertia in the momentum equation, leads to a simplified mathematical representation, expressed as equation 1.

The application of this detailed hydraulic method would be restricted to small areas only. Therefore, it only served as a benchmark for a further simplified two-dimensional approach where Manning's equation provides a closed-form expression for the two-dimensional velocity vector (u,v) that is used in the momentum equation 1

$$\frac{\partial h}{\partial t} + \frac{\partial uh}{\partial x} + \frac{\partial vh}{\partial y} = S_p \tag{1}$$

t   time variable,
x, y   spatial variable,
h   water level,
u   depth averaged velocity in x direction,
v   depth averaged velocity in y direction,
S_p sink/source term as the exchange value with the pipe flow model,

The RisoSurf approach is described in greater detail in Ettrich et al. (2004).

From a mathematical point of view, the new and crucial point in this approach of hydraulic surface flow simulation is the coupling of the shallow water equation model of surface flow with the dynamic sewer flow model. Sink/source term S_p, being the primary term of exchange with the pipe flow model, requires particular consideration of numeric stability. This crucial point of coupled hydraulic flow routing procedure is further discussed in the section "Coupling Modules RisoSurf and HamokaRis" below.

Dynamic Sewer Flow Modeling

Sewer flow is simulated by applying fully dynamic flow routing of unsteady, gradually varied flow and solving Saint-Venant Equations numerically in an explicit difference scheme. The explicit difference scheme is applied in variable time steps that are permanently adjusted to the COURANT-criterion, guaranteeing numerical stability (Schmitt 1986).

At each time step, the procedure of dynamic flow routing starts by computing flow values for each conduit (sewer segment between nodes) based upon the momentum equation and instantaneous water levels at the nodes at the end of the last time step. In the next step of the dynamic flow routing procedure the flow volume is balanced at each node, taking into account inlets from house drains and all surface inlets connected as well as inflows and outflows from sewers connected at the nodes. The resulting change of volume is drawn to free water surface "available" at the node, thus producing a change of water level at the node. In order to improve numerical stability, the two phases are applied in a half-step/full-step procedure during each time step as described in (Roesner et al. 1988; Schmitt 1986).

The underground sewer system is represented by a network of nodes and conduits (sewer segment between nodes). In contrast to conventional modeling, not only manholes but also street inlets and house drains are considered as extra nodes to fully achieve the connection of surface and underground drainage system at all locations where interaction between surface and sewer flow and potential flooding might occur.

Modeling the Interaction of Surface and Sewer Flow

The simulation of the interaction between surface and sewer flow is based upon the definition of exchange locations. Each runoff area is allocated to one specified exchange location as illustrated in Fig. 5. Here, all relevant information for surface and sewer flow simulation (instantaneous runoff, water level, and exchange volume) is available at the beginning of each time step for all simulation modules and is renewed at the end of the time step in the following way:

1. The hydrologic runoff model RisoReff only supports uni-directional flow and is applied to all areas not considered for surface flow. Computed runoff from those

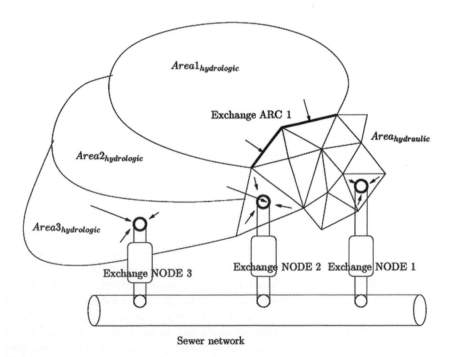

Fig. 5 Possible links and interaction of surface runoff, surface flow and sewer flow at exchange nodes

"hydrologic areas" is passed to the single exchange location to which the area is connected. The exchange volume would be the runoff volume in the according time step.

2. Areas simulated with the hydrologic model approach can be connected to the underground drainage system in two alternative ways:
   - "hydrologic areas" directly discharging to the sewer system via surface inlets or private drains (this procedure works much like the well-known approaches of commercial urban runoff simulation software of the EXTRAN type. If the instantaneous sewer capacity at the exchange location were not sufficient to take in the surface runoff, or if water escapes from the sewer system due to surcharge, the water volume would be stored at the specific exchange location in an artificial storage and re-entered to the sewer system with diminishing surcharge)
   - "hydrologic areas" discharging to surface areas where surface flow is considered by the hydraulic model RisoSurf (such an exchange location can be seen as an edge where the inflow to the "hydraulic area" would be evenly distributed over the full length of the edge ["exchange arc 1" in Fig. 5]).

3. The hydraulic surface flow module RisoSurf allows bi-directional exchange of runoff volume:
   - from the surface area to the sewer system – if there is sufficient sewer capacity
   - from the sewer to the hydraulic surface in case of sewer surcharge when the water level in the sewer system rises above ground level

4. In case of surcharge, the water level above ground as provided by surface flow simulation module RisoSurf at exchange nodes would be used by dynamic sewer flow module HamokaRis in the momentum equation in the following time step: if the balance of flow volume at the nodes in HamokaRis results in a water level above ground, the associated surplus volume would be transferred to the surface flow simulation by "storing" this volume in the exchange location and taken into account by RisoSurf in the sink/source term S_p (equation 1) the next time step.

Coupling of Modules RisoSurf and HamokaRis

The implementation of coupled hydraulic flow routing for surface and sewer flow modules RisoSurf and HamokaRis requires particular consideration of numeric stability and observation of continuity as well. Numeric stability has been secured by a synchronized administration of dynamic time step selection according to Fig. 6.

The Engine defines the instantaneous time step size as a minimum of instantaneously allowable time increments delivered by the single modules. Continuity of mass has been observed by creating an exchange table where all instantaneous inflows, outflows, and storage volumes at any exchange location are balanced during the

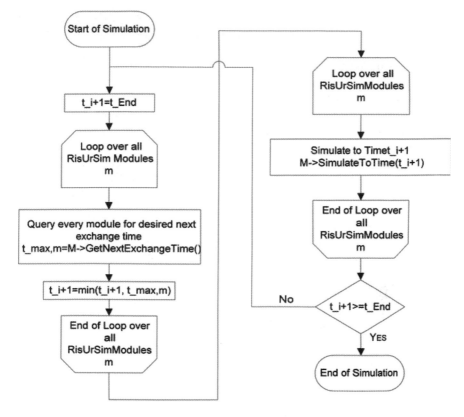

Fig. 6 Procedure to synchronize time steps of simulation modules *RisoSurf* and *HamokaRis* in simulation tool *RisUrSim*

simulation and verified in a system-wide total balance at the end of simulation period. The overall model structure of RisUrSim, including data processing and interaction and synchronization of simulation modules, is shown in Fig. 7.

Model Application – Case Study

One of the test areas used to prove the concept of the RisUrSim method is a sub-catchment in the city of Kaiserslautern (Fig. 8). Some of the houses in the southern street of the test area have been subject to basement flooding during heavy rainfall in the past. To prepare the model application, detailed surveying has been carried out to accurately describe flow-relevant surface areas. In addition, a flow monitoring device has been installed to gather data during rainfall events for model calibration under surcharge conditions. This, however, has not been successful, as during the period of monitoring not a single surcharge or even flooding event occurred.

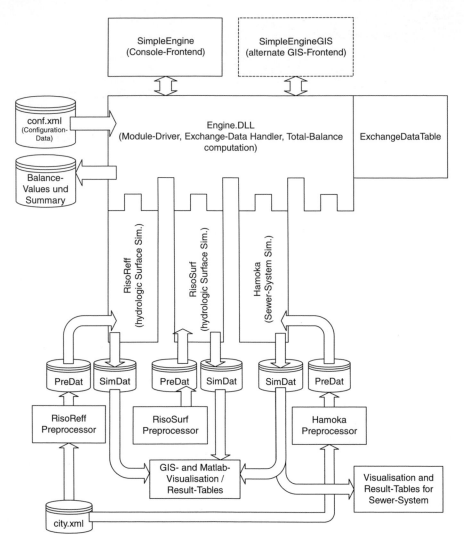

Fig. 7 Overall structure of the RisUrSim simulation tool ("Engine")

Data Needs

Data needs for detailed simulation of urban flooding are related to a distinct representation of the runoff areas, the surface characteristics and local constraints that possibly influence surface flow patterns, and the underground sewer system itself.

a. Runoff areas. In order to most accurately describe their runoff behavior, single runoff areas within the sub-catchment have been classified and sized as shown in Table 1. Distinct model parameters have been applied to quantify interception, depression storage, and infiltration.

Fig. 8 Test case area Kaiserslautern-Erzhuetten

Table 1 Classification of surface areas within the catchment

(1) imperviously paved areas	(2) previously paved areas	(3) unpaved areas
1.1 streets and other traffic areas	2.1 streets and other traffic areas	3.1 green roofs
1.2 span-roofs (>10 % grade)	2.2 yards, private parking lots, etc.	3.2 lawns, garden area
1.3 flat roofs (except green roofs)		3.3 garden area
1.4 yards, private parking lots etc.		

b. Surface characteristics. A Digital Terrain Model (DGM) has been set up in a high resolution in order to apply detailed hydraulic surface flow simulation module RisoSurf. The DGM includes distinct levels of street cross-sections, location and height of manholes and street inlets, side-walks and street-curbs as well as the borderline between public (street, side-walk) and private space (building sites). Fig. 9 shows the mathematical representation of the street surface area in a triangular network for the hydraulic surface flow module RisoSurf. Adjustments had to be made with regard to the representation of street-curbs by a double polygon describing the location and different heights of street gutter and side-walk. For practical reasons it was decided to exclude areas on private sites from surface flow simulation in regard to the lack of detailed physical surface data.

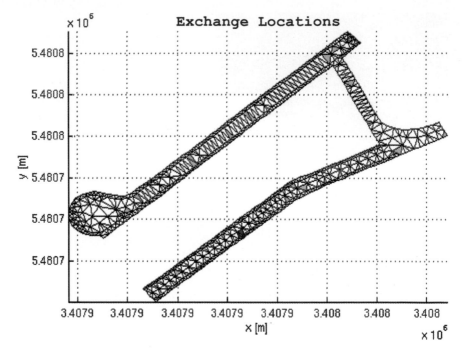

Fig. 9 Mathematical representation of the street surface in a triangular network in surface flow module RisoSurf

c. Detailed sewer system. The sewer network, normally represented by manholes nodes and sewers only, has been further detailed by
   • including private house drains represented as closed pipes uni-directional input node, and
   • including street inlets represented by a bi-directional exchange location and followed by a closed pipe.

These inlet elements have been connected to the public sewer system either via existing adjacent manholes or by introducing extra system nodes ("virtual manholes"). This is done automatically by a software routine depending on the location of the single inlets in relation to the sewer line.

Simulation Results

Due to the fact that no surcharge or flooding event could be monitored, the RisUrSim software has been applied to a variety of test scenarios using synthetic design storms. These applications have been done to verify the most crucial model features of hydraulic surface flow simulation and the interaction between surface flow and sewer flow under surcharge and flooding conditions. Test simulations have been done with single modules using simple structures as input data, because data from "real-world" systems would result in systems too complex for testing purposes. Since the demand on data to achieve realistic simulation results, especially to model the surface structure, is very high compared to "traditional" hydrologic surface representation in sewer-system simulation, even the real-world tests can cover only small regions.

The simulation results of the real-case system Erzhuetten are shown in Fig. 10 in terms of water level distribution along the street surface at 15 min and 25 min

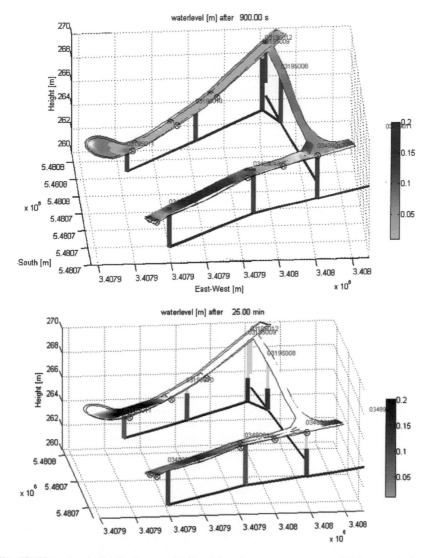

Fig. 10 Water level distribution as simulated for the test case system Erzhuetten during a synthetical design storm after 15 and 25 minutes

simulation time, respectively. In this system the surface elevation decreases from north-east (right) to south-west (left) while the sewer-system flow direction is oriented in the opposite direction. This has led to problems with flooding in this area in the past. The representation of simulated water levels in the manholes and on the street surface illustrates the surface flow pattern from surcharged, flooded manholes to street areas with lower surface levels (on the left side of the graph). This proves that with the RisUrSim Software the surface flooding could be reproduced realistically.

Conclusions

European Standard EN 752 triggers more intense consideration of the flooding phenomenon in urban drain flow modeling. In the RisUrSim approach particular recognition is given to detailed surface flow simulation and the interaction between surface and sewer flow during times of surcharged sewers. This approach has been earlier denoted as "dual-drainage." In the context of research project using RisUrSim, the following appreciations can be stated:

- Simulation of flooding in urban drainage systems requires some kind of hydraulic surface flow being permanently linked with sewer flow in the flow routing procedure in order to more accurately assess water levels and derive possible damage of flooding.
- The coupling of hydraulic surface flow modeling with dynamic sewer flow routing in flooded areas requires specific consideration to proper synchronization of time step selection and mass conservation at exchange locations.
- Hydraulic surface flow simulation requires detailed physical representation of surface areas within the drainage system in order to accurately describe flow patterns in case of surface flooding. Such detailed data is not generally available and will restrict practical application in the near future. This data need, however, should trigger further GIS development and applications.

Acknowledgements The RisUrSim research project has been funded by the German Ministry of Education and Research (BMBF) and the Research Council of Norway within the EUREKA framework (Σ!2255 RisUrSim). Besides, the author is greatly indebted to the project partners Fraunhofer-ITWM, the cities of Kaiserslautern and Trondheim as well as the insurance company Deutsche Rück. Special thanks are given to the project partners in Norway (Prof. Wolfgang Schilling, Dr. Sveinung Sægrov) and the research co-workers Christoph Garth and Michael Hilden.

References

Akan AO, and Houghtalen RJ (2003) Urban hydrology, hydraulics and stormwater quality. John Wiley, Hoboken, N.J.
Ashley RM, Hvitved-Jacobsen T, Bertrand-Krajewski JL (1999) Quo vadis sewer process modeling. *Water Science Technology*. Vol. 39, 9:9–22

DWA (2006) Hydraulische Berechnung und Nachweis von Entwässerungssystemen (Hydraulic calculation and verification of drainage systems). Arbeitsblatt A 118, DWA-Regelwerk, Hennef, Germany

CEN (1996, 1997) Drain and sewer systems outside buildings – Part 2: performance requirements (1996); Part 4: hydraulic design and environmental considerations (1997). European Standard, European Committee for Standardization CEN, Brussels, Belgium

Ettrich N, Steiner K, Schmitt TG, Thomas M, Rothe R (2004) Surface models for coupled modeling of runoff and sewer flow in urban areas. Conference paper presented at Urban Drainage Modeling 2004, Dresden, Germany

Djordjevic S, Prodanovic D, Maksimovic C (1999) An approach to simulation of dual drainage. Water Science and Technology, Vol. 39, 9:95–103

Hilden M (2003) Extensions of shallow water equations. Ph.D. thesis, Dept. of Mathematics, Kaiserslautern University

Roesner LA Aldrich JA, Dickinson RE (1988) Storm water management model user's manual, version 4: addendum I, EXTRAN, EPA/600/3-88/001b (NTIS PB88236658/AS), Environmental Protection Agency, Athens, GA

Schmitt TG (1986) An efficient method for dynamic flow routing in storm sewers. In: Proc. of the international symposium on urban drainage modeling. Dubrovnik, Yugoslavia, pp. 159–169

Schmitt TG (2001) Evaluating hydraulic performance of sewer systems according to European Standard EN 752. Water 21, Magazine of the International Water Association February 2001, pp. 29–32

Schmitt TG, Thomas M, Ettrich N (2004) Analysis and modeling of flooding in urban drainage systems. J. of Hydrology, Special Issue Urban Hydrology, Vol. 299, 3–4:300–311

Schmitt TG, Thomas M, Ettrich N (2005) Assessment of urban flooding by dual drainage simulation model RisUrSim. Water Science & Technology, Vol. 52, 5:257–264

Environmental Noise

Martin Rumberg

Overview

As an introduction to environmental noise issues, some of the sources and the major effects are defined. Then the question of sound propagation and noise mapping is addressed, followed by a discussion on the assessment of noise exposure as a central question in what is to be done out of acoustical data in a spatial context. The article ends with some conclusions about the state-of-the art and future challenges.

Introduction

Environmental noise may be defined as unwanted sound that is caused by emissions from traffic (roads, air traffic corridors, and railways), industrial sites and recreational infrastructures, which may cause both annoyance and damage to health. The area of environmental noise is subject to two important restrictions. The first is that spatial systems or infrastructures are to be assessed, not the sources of sound emissions like cars or aircrafts. Therefore, environmental noise assessment and mitigation is not concerned with technical questions, such as constructing cars in a way that makes them quieter, but rather with the question of infrastructure planning and its impact on residential and leisure areas.

The second major restriction is that there are some kinds of noise that may not be assessed in the environmental context – the most important one being neighborhood noise, which is a question of architecture, of social behavior, and of legislation, but not of spatial planning – although there are some connections to planning.

M. Rumberg
Technische Universität Kaiserslautern, Lehrstuhl Stadtplanung, Pfaffenbergstr.95,
67663 Kaiserslautern, Germany, E-mail: rumberg@rhrk.uni-kl.de

G. Steinebach et al. (eds.), *Visualizing Sustainable Planning*,
© Springer-Verlag Berlin Heidelberg 2009

Noise Effects on Health

Noise in the environment or community seriously affects people, interfering with daily activities at school, work and home and during leisure time. The World Health Organization identifies the following as the main health risks of noise:

- annoyance is the most common effect of environmental noise (annoyance can be considered as either a health effect or social impact of noise depending on the definition of "health")
- hearing impairment, including tinnitus (hearing loss is an obvious effect of noise but it is not is major concern with respect to environmental noise because normally we do not find sound power levels high enough to cause hearing loss)
- interferences with social behavior (aggressiveness and helplessness)
- interference with speech communication
- sleep disturbance and all of its consequences on a long- and short-term basis
- cardiovascular effects (an increase in the risk of heart attacks)
- hormonal responses (stress hormones)
- decreases in performance at work or school

Environmental noise may be considered one of the main local environmental problems in Europe and America. Furthermore environmental noise is causing an increasing number of complaints from the public. However, action to reduce environmental noise has had a lower priority than that taken to address other environmental problems, such as air and water pollution.

However it has been estimated that around 20% of the European Union's population or, close to 80 million people, suffer from noise levels that are considered unacceptable – meaning that many or most people become annoyed, that sleep is disturbed, and that serious long-term health effects can be expected. The European Union calls these "black areas." An additional 170 million citizens are living in so-called "grey areas" – these are areas where the noise levels are such to cause serious annoyance during the daytime – but no serious health damages. Only 150 million Europeans live in "white areas" that are not seriously affected by environmental noise. These are mostly rural areas or privileged places in urban areas.

Noise Mapping and Indicators for Environmental Noise

The data available on noise exposure is generally poor in comparison to that collected to measure other environmental problems and noise data is often difficult to compare due to very different measurement and assessment methods.

To effectively cope with noise problems, planners need a quantitative and informative basis for discussion. A key instrument is noise mapping. Noise mapping is the representation of acoustic data in a cartographical format and the assessment of noise exposure in the context of public health and quality-of-life. So it is a multi-level modeling process. Its benefits are the visualization of acoustic data, and the correlation of different datasets by combining the information with a geographical information system (GIS). Thus population density can be related to noise exposure to generate information about the number of people exposed to noise levels above a certain criteria level. In general the purpose of noise mapping is

- to give an accurate statement of sound levels in a location,
- to provide noise trend data,
- to establish exposure levels of a population for risk estimation purposes,
- to identify pollution hotspots and relatively quiet areas for the definition of priorities,
- to yield information as to the effectiveness of management schemes, and
- to inform the public as decision makers.

Noise maps and noise risk estimations are based on specific acoustical descriptors and indicators. The properties of noise that are important in the environment are sound power, frequency (higher frequencies cause a different effect than low frequencies), and time distribution (for instance nighttime noise is a very different issue from daytime noise because of different locations and activities).

The decibel unit is used in acoustics to quantify sound levels relative to a 0 dB reference. This reference may be defined as the sound pressure at the threshold of human hearing, which is conventionally taken to be 20 micropascals. The reason for using the decibel is that the ear is capable of detecting a very large range of sound pressures. The ratio of the sound energy that causes permanent damage from short exposure to the limit that (undamaged) ears can hear is above a trillion. To deal with such a range, logarithmic units are useful: the log of a trillion is 12, and this ratio represents a difference of 120 dB. It was discovered about a hundred years ago that our perception of loudness is roughly logarithmic In other words, you have to multiply the sound energy by the same factor to have the same increase in loudness.

As sound pressure levels are based on a logarithmic scale they cannot be added or subtracted in the usual way. If one machine is emitting a sound level of 60 dB, and a second identical machine is placed beside the first, the combined sound level is 63 dB, not 120 dB. Instantaneous sound levels above 85 dB are considered harmful, while 120 dB is unsafe and 130 dB causes physical damage to the human body. When the difference between two noise levels is 10 dB(A) or more, the amount to be added to the higher noise level is zero. In such cases, no adjustment factor is needed because adding in the contribution of the lower in the total noise level makes no perceptible difference in what people can hear or measure. For example if the noise level is 65 dB and you add another source that produces 50 dB, the noise level will still be 65 – a question of sound masking that is very important for planning. One could say that in combined noise situations there is no use in reducing the noise of sources with lower sound power.

The sensitivity of the human ear to sound depends on the frequency of the sound. People hear some frequencies better than others. If a person hears two sounds of the same sound pressure but different frequencies, one sound may appear louder than the other. This occurs because people hear high frequency noise much better than low frequency noise. Noise measurement readings can be adjusted to correspond to this peculiarity of human hearing. An A-weighting filter that is built into the instrument de-emphasizes low frequencies. Decibels measured using this filter are A-weighted and are called dB(A). Legislation on environmental noise normally gives exposure limits in dB(A). A-weighting serves two important purposes:

1. It gives a single number noise level by integrating sound levels at all frequencies.
2. It gives a scale for noise level as experienced or perceived by the human ear.

Environmental noise in general is not constant over time but shows a complex time structure. So the instantaneous sound level as short-time descriptor is not a sufficient description for environmental noise. There is a need to look for long-term descriptors such as average sound levels, weighted average sound levels (such as LDEN, a 24-h hour average level in which evening and nighttime is weighted), peak (or maximum) sound levels, and NATs (number above a defined threshold, which is especially interesting during the nighttime or for communication disturbance).

Generally noise data for noise maps may be achieved by two methods: measurement and prediction. The current trend is to use modeling based on complex propagation algorithms to predict noise data because measurement is very expensive and time-consuming and is not suitable for future situations. The main problem of noise prediction is that the input models should not be too sophisticated but manageable and predictable. However, without including factors it becomes difficult to achieve measures. For example, fluent traffic with moderate and homogenous speeds is less noisy than stop-and-go traffic; however, traffic flow is very hard to predict in advance and so it is no simple problem to work with accurate data in this case. As a result, there is a need to look for suitable estimations.

Modeling consists of entering sound power levels to represent each significant source of noise associated with a given environment and then calculating the total sound pressure level that results from all these sources acting together at one or more distant points of interest. Sound propagation is a very complex modeling task with respect to many influencing factors. The calculations from source input level to receptor output level require the consideration of constant and fluctuating factors, such as:

- source type (point, area, line, etc.)
- distance loss
- atmospheric air absorption losses
- ground interaction effects
- blockage from other on-site structures or equipment
- losses from intervening trees or other off-site obstructions
- losses from noise barriers or berms
- potential losses due to intervening terrain
- wind direction and speed
- atmospheric temperature gradients
- cloud cover

All of these influences on sound propagation are considered in every model we run using algorithms from accepted international standards and research, such as ISO 9613 (Sound Propagation Outdoors) and Verein Deutscher Ingenieure (VDI) 2714 (Schallausbreitung im Freien).

A very simple example for noise propagation modeling is distance loss for a point-source without considering noise barriers or atmospheric aspects. Behind it there is the simple physical model of energy conservation – since the power level is related to the spherical area, it is not a continuous function but a quadratic equation.

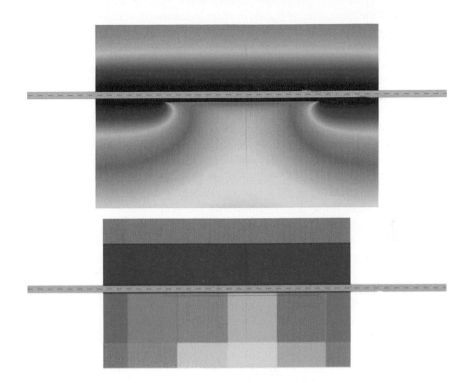

Distance loss per meter nearby the source is much higher than in a greater distance. Losses from barriers are very important for accurate noise mapping in settlement structures. The effect caused by buildings, topography, and special noise barriers such as walls is very significant – up to 35 dB(A) but has to be predicted very precisely because of wave effects on top and at the sides of the barrier.

The estimations of a barrier effect for a roadside situation shown above are both modeled with an ISO 9613 algorithm but with a different level of detail, specifically a different raster scale. The first one is calculated in a 1-m grid and gives a realistic impression of the exposure level behind the barrier – the emission in the middle of the barrier is well below the emissions at the ends. The second estimation shows the same situation but calculated with a 50-m grid. It is obvious that the effects cannot be estimated properly this way, demonstrating the need to work with a raster scale that is much smaller than for other environmental modeling purposes.

Reflexion is a phenomenon that has to be generalized because of the complexity of the environmental situation. And it is possible to generalize it because of a relatively small influence on the sound level.

A Practical Approach to Noise Mapping:
Kaiserslautern City

As an example of environmental noise assessment a road-traffic noise map for the western part of the city center of Kaiserslautern is discussed. It shows an estimation for an A-weighted long-term 24h-sound level - the level day/evening/night (which is a 24-hour average sound level with an addition of 5 dB(A) for noise in evening time and an addition of 10 dB(A) in the night time. This is a European standard indicator for so-called general annoyance. DEN-levels from about 75 to 80 dB(A) are situated directly at the roadside down to about 35 dB(A) inside the courtyards. This effect is mainly caused by noise barrier effects of the buildings.

There is a high influence of building structures as noise barriers in this high-density-area. One benefit of accurate noise mapping is to exploit the spatial noise level variations. Even though many dwellings are exposed to Lden = 65–70 dB, outdoor levels in the range of 35–40 Lden are not rare in the cities, and classical courtyards in European cities are a typical reason for this phenomenon. The EU directive on ambient noise puts emphasis on the protection of quietness. Quiet areas and "quiet facades" shall be shown on noise maps. The concept of a relatively quiet facade has been introduced. One questions to consider is what are the effects on the annoyance and sleep disturbance when dwellings have one noisy and one quiet side?

To work out quantitative modeling from this data it is necessary to integrate more spatial data, depending upon what is to be modeled. Land-use designation (such as residential area) is a very common indicator for target-actual-comparison – may be

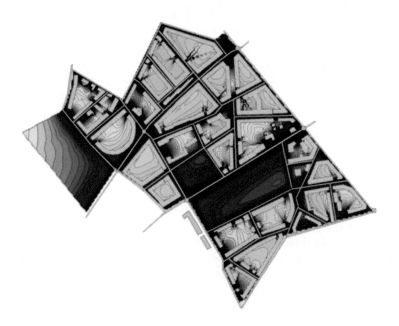

55 dB(A) as target. Population data is important for risk estimation and for the assessment of the effectiveness of noise mitigation and management schemes. Socio-economic indicators like household income are not commonly integrated but may be feasible for the specific quantitative analysis of noise effects in a socio-economic dimension (for example, the effects of poor environmental quality on social segregation). The suitable level of detail for these modeling tasks includes dwellings, buildings, or blocks.

Based on this data, common dose-response relationships curves for annoyance. It shows the estimated percentage of people annoyed by noise (road noise is the curve in the middle) is proportional to the day/evening/night level outside the most exposed facade of the dwelling. It looks precise but actually it is a very rough estimation. For instance, it does not consider the effect of a very quiet second facade on the annoyance level. But better, more detailed dose-response relationships were not available until now because of poor empirical data.

Dose-response relationships between noise level and annoyance show variations between different noise sources. Aircraft noise is much more annoying than road traffic noise, while railroad noise is less annoying. Thus, if a location is affected by noise from different sources – e.g., road traffic and aircrafts as a typical combination - the summation is a very sophisticated process, both mathematical and technical. The result is a rough annoyance risk estimation per dwelling.

The benefit of this modeling technique is to generate an overview of noise exposure and representative effects that is not too sophisticated and is manageable for optimization and planning in a wider area.

Another approach to modeling noise effects is to look at the acoustical quality of areas for activities – especially activities a site is designed for. The picture shown here typifies a recreational zone in a residential area (this is an example from Berlin) with balconies, park benches, and a playground. Suitable acoustic quality criteria are moderate annoyance (otherwise recreation would not be possible), gentle conversation (speech clarity at 1 m), and natural sounds.

These criteria represent three different aspects of an acoustical impact associated with the activity of recreation in a three-level assessment. An ideal relaxation area is one in which the natural soundscape dominates. As nature is not quiet, it is not necessary to have 0 dB(A). 40 dB(A) as a basic sound level throughout the day is sufficient for relaxing natural soundscape, although listeners might sometimes hear a car or a train. So there is no benefit in setting lower limits than 40 dB(A).

An area is to be considered as suitable for recreation when the constant (or near constant) sound level allows undisturbed communication without getting words wrong or having to shout. It is a bit tricky to define this level because speech disturbance is strongly affected by peak levels. The peak level should not exceed 55 dB(A) (that is 10 dB(A) less than the speech level itself. With 47 dB(A) we can be sure that these maximum levels do not occur more than 5% of time. Minimum requirements may be seen on a level that starts to cause annoyance to many people. In relaxation areas that could be a 10% level for moderate annoyance, and while coping with traffic noise that is about 52 dB(A).

Looking back to the example of Kaiserslautern city there are many courtyards with this "gold standard" – absolutely suitable for relaxation in spite of being situated in a high-density and high-traffic zone.

Challenges

A main challenge for the near future is to increase the efficiency of noise mapping by standardizing processes and bundling tasks (GIS-based modeling). In Europe, this will be highly influenced by the implementation of the European Union's directive on environmental noise that will standardize some of the methods European-wide.

Another task will be the improvement of dose-effect relationships on a higher level of detail (e.g., quiet facades / insulation) – as the state-of-the-art is not suitable for prediction purposes.

With the systematic integration of noise issues into multi-criteria decision support systems for urban planning, dealing with questions of settlement, traffic, and environment may become possible in a few years.

Work-Facilitating Information Visualization Techniques for Complex Wastewater Systems

Achim Ebert and Katja Einsfeld

Introduction

The design and the operation of urban drainage systems and wastewater treatment plants (WWTP) have become increasingly complex. This complexity is due to increased requirements concerning process technology, technical, environmental, economical, and occupational safety aspects. The plant operator has access not only to some timeworn filers and measured parameters but also to numerous on-line and off-line parameters that characterize the current state of the plant in detail. Moreover, expert databases and specific support pages of plant manufactures are accessible through the World Wide Web. Thus, the operator is overwhelmed with predominantly unstructured data. He is overstrained with the task of finding relevant information, spotting important trends, and detecting optimization potential. In the worst case the operator is overloaded with irrelevant information and, consequently, is no longer able to respond adequately to exceptional circumstances.

The first part of our contribution describes the project SIMLAR, which covered information visualization approaches for various tasks and devices. Visualizations for mobile devices are especially useful in the case of large, distributed wastewater systems. Especially in the case of an alarm, mobile process information can help to determine if it is necessary to visit the plant.

The second part details the project KOMPLETT, which researches intuitive visualization systems for complex plants. Considering the fact that the plant is unmanned and thus even non-experts have to cope with the system, it has to meet diverse requirements: it has to visually and intuitively indicate trends and irregular behavior of quantitative data. Depending on the context, it has to show semantic relations between currently relevant qualitative data like the technical process or the physical layout of the plant and textual data like plant documentation.

A. Ebert(✉)
Technische Universität Kaiserslautern, Fachbereich Informatik, HCI & Visualization Lab, Postfach 3049, 67653 Kaiserslautern, Germany, E-mail: ebert@cs.uni-kl.de

G. Steinebach et al. (eds.), *Visualizing Sustainable Planning*,
© Springer-Verlag Berlin Heidelberg 2009

SIMILAR

To achieve the goal of device-sensitive mobility, it is necessary to transform the available data into usable, non-overwhelming information, which is available at any location at any time. Thus, a demonstrator has been developed to visualize the information on small mobile devices like mobile phones, Pocket PCs or Tablet PCs. These instruments can be easily carried by the staff of a wastewater treatment plant.

Related work can be found in Goose et al. (2003) on mobile maintenance, in Lipman (2002) on 3D steelwork models on a mobile, and in Schilling and Coors (2003) on 3D maps.

System Framework

The SIMILAR system framework (Fig. 1) is built out of two major parts: a data layer and a presentation layer. The data layer is implemented as a relational database containing all information about the WWTP as well as references to external media files (images, videos, and 3D data). The presentation layer is responsible for generating dynamic pages with an appropriate layout. The Tomcat Java servlet container is used for the presentation layer. Tomcat can generate dynamic pages by processing JSP (Java Server Pages) tags that are embedded in page templates.

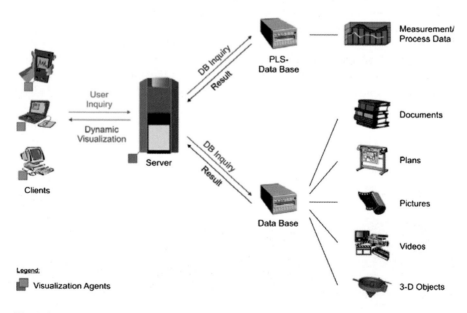

Fig. 1 SIMILAR system framework

The single, dynamic generated pages are encoded in HTML, which enables the system to be accessed by any internet browser. When a browser is launched and accesses the server for the first time, a visualization agent collects the needed information about the computational and graphical capabilities of the (mobile) client device (Ebert et al. 2001). The server stores this data for the duration of the session. By comparing the acquired values with the information stored in an internal knowledge base, the visualization agent is able to compute device-adjusted adequate visualizations.

3D Visualization

In order to provide a flexible way for a client-independent 3D visualization, we have implemented three different solutions that are automatically chosen by a visualization agent depending on client-capabilities during runtime: VRML viewer, 3D server plus Java applet, and 3D servlet. VRML allows a high degree of interactivity, but requires sufficient computing and 3D graphics power. The 3D server approach a Java applet must be started on the client that monitors the user interactions with the client. The applet connects to the server, which renders the actual view based on the measured interaction parameters and then sends the resulting image back to the client. This requires high network and server capabilities. Compared to the VRML-approach the degree of interactivity is significantly lowered. The 3D servlet-approach only requires a web browser. The client sends standard HTTP requests to the server, which capsulate the actual user input. The servlet then renders the image (e.g., in PNG format) and sends it back to the client. There are low requirements for client, server, and network, but results in a very low degree of interactivity. Figure 2 shows two examples of device sensitive visualization for a desktop system and Pocket PC.

Based on the client's system and the server properties the visualization agents decide which solution will fit best for the actual case. For example, on a desktop system and – in the case of a small number of triangles – for a PDA, sending the

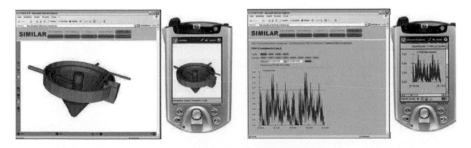

Fig. 2 3D model of a tank and process monitoring and visualization

VRML file is a good solution. For a higher number of triangles, the 3D server method is an appropriate solution for PDAs. If Java is not available on the PDA, or for slow network connections, the 3D servlet solution will be chosen by the agents.

KOMPLETT

In contrast to SIMILAR, the KOMPLETT project ("komplett" is German for "complete") is still a work in progress and currently more focused on the development of intuitive metaphors for information rich environments than on device sensitivity. The goal of the project is the development of a small, unmanned Waste Water Treatment Plant (WWTP) with two distinct water circles and resource recycling ambitions. This requires an intuitive representation of large amounts of heterogeneous data (e.g., process data, optimization suggestions, expert knowledge, maintenance instructions, and interrelations between them) depending on the user profile. The visualization should provide detailed and revealing information for the remote expert as well as intuitive access to relevant instructions for the non-expert on site. In this context multiple semantic perspectives are essential.

The presented 3D interface is called HANNAH, which is short for "Here And Now, Near At Hand." The name accounts for the claim of creating an HCI-framework that is context sensitive, adaptable, ergonomic, intuitive, vivid, allows direct manipulation, and provides direct animated feedback.

System Framework

In order to develop a system that instantly adapts to changing user-, task-, and device contexts and that allows the user to intuitively configure the interface depending on his or her needs by simply plugging together virtual 3D components, the system architecture has to be highly modular. The needed 3D support is made available by the use of the graphics library OpenGL. The modular and hierarchical architecture allows the system to consecutively develop more complex visualization objects in multiple levels of abstraction, handle them consistently, and plug them together easily: individual VisObjects or CompoundVisObjects can be rotated or shaken (ShakeMetaphor) to offer motion parallax depth cues (Ware and Frank 1996). They can be moved along animated paths and they can be moved in and out of view (Structure on Demand concept) and scaled depending on the current focus (TacticalZoom).

Databases and Ontology

The data resulting from measurements on the plant is stored in databases. However, the goal of the KOMPLETT-project is not only to visualize process data and static process views, but also to efficiently, intuitively, and visually organize more information about the process and to interconnect these heterogeneous information items. For this purpose, we decided to collect the information as ontology. The ontology modeled in Protégé contains process steps, the process sequence, reactors, measuring instruments, links to diverse information material, table names of measured data, many other information items, and the semantic links between all these items.

Process Data Visualization

Most conventional process data visualizations only consist of simple two-color line graphs for one parameter in a specific time interval. Considering the findings of Information Visualization research concerning perception, visual scales, and intuitive metaphors, our approach strives to provide the user with more efficient visual analysis tools.

Figure 3 adds the visual scale color to the concept of simple graphs. The value of the process parameter is mapped to y-position as well as to a continuous color gradient. The color gradient can be interrupted at one or more points (in this case between cyan and green and between orange and red) to represent that the value is below or above a certain limit. The small sphere on the diagram is moved following the mouse pointer and continuously displays the exact value under the sphere.

When more than one parameter has to be displayed or the time interval is enlarged, the user usually has the option of choosing between time multiplexing (displaying one diagram after the other) or space multiplexing (all diagrams are scaled down and displayed simultaneously, see Fig. 4a). Thus, the user has to decide whether he or she wants to forego the possibility of comparing diagrams in order

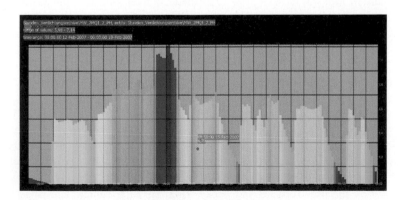

Fig. 3 Process data in a single diagram

Fig. 4 Process data: (**a**) Space multiplexing (**b**) ColorPlane

Fig. 5 Color rolls

to perceive details. The idea of representing a large series of values with few color-coded pixels, the pixel-based techniques, has a long history in information visualization research. E.g., Ankerst et al. (1996) proposed CircleSegments and Kincaid and Lam (2006) developed the Line Graph Explorer. In our system, we use ColorPlanes to visualize the value of one parameter in multiple time intervals. The ColorPlane in Fig. 4b shows temperature measurements in our office. Each row represents one day. Comparing the rows easily reveals certain patterns: At about 9:00 a.m., the temperature falls due to the opening of the windows. At about 1:00 p.m., the temperature rises as the door is closed during lunchtime.

Making use of the third dimension, multiple ColorPlanes with data from different process parameters can be visualized as ColorRolls (Fig. 5). The rolls can be simultaneously rotated in order to move the time interval of interest to the front. Due to perspective projection, a natural focus and context effect is created. Weber et al. (2001) proposed a similar approach for visualizing time-series on spirals.

Another new 3D process data visualization metaphor that serves as a good compromise in the time vs. space multiplexing dilemma is the Rotary Diagram shown in Fig. 6. Similar to the cards in a rotary file known from daily life, the diagrams can be rotated by the user forward and backward around the center. The rings in the center (Fig. 6b) are not only used to enforce the rotary file metaphor. They visually

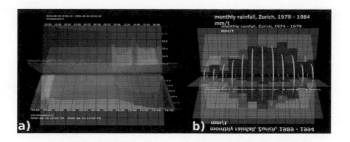

Fig. 6 Rotary Diagram (**a**), RotaryDiagram with AverageRings (**b**)

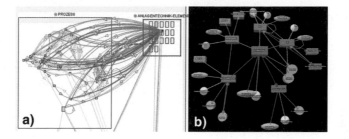

Fig. 7 Process ontology in Protégé (**a**), ConeView of process in HANNAH (**b**)

indicate the average value of each column from all diagrams. These average value rings help to spot irregularities in the rotated diagrams. Moreover, the semitransparent style of the diagrams allows the user to directly compare consecutive diagrams.

Process Visualization

As explained before, the sequence of process steps needed for the operation of the plant, the reactors involved, the data measured, and other information are organized in an ontology. Thus, one way to visualize the functionality of the plant is to visualize parts of the ontology. Fig. 7a shows a cluttered ontology visualization in Protégé.

In the ontology visualization of HANNAH the diverse semantic categories (e.g., process steps, reactors, process data) are displayed in diverse colors, shapes, and at diverse depth levels. The relationships among the process steps that are currently in focus are emphasized by animated arrows. Fig. 7b shows the ConeView ontology visualization. It can be used when the user wants to focus on a particular aspect of the process. The selected focus element is displayed in the center and front. Process elements that are directly connected to the focus element are rendered in a first circle around the focus point. This circle is more distant to the user than the focus point. The second circle, which is even more distant, contains items

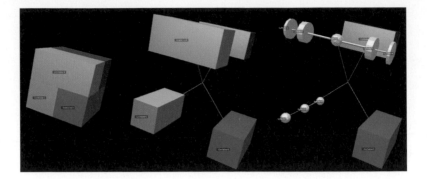

Fig. 8 Three states in animated explosion view

that are connected to those of the first circle. When the user selects another item, it is moved to the focus position and its connected process elements are organized around it in an animated way.

Process Plant Visualization

Extending the 2D explosion-view metaphor to 3D, we have created a visualization tool for the physical layout of the plant that allows for exploring it hierarchically. Fig. 8 shows three states in the exploration of the example geometry. The 3D ExplosionView tool is implemented by creating a containment hierarchy of elements, for each hierarchy level a graph representing the way in which the elements are connected to each other, and for each element the direction in which it is expected to explode. With this data, the geometry can be exploded step by step or automatically to reveal one specific element (e.g., a reactor that has to be inspected). A similar approach has been used for assembly instructions by Agrawala et al. (2003).

Interconnected Perspectives

With the help of the previously described process ontology, we are able to visually relate information items from diverse categories or perspectives to each other. E.g., process data can be related to the reactor where it is measured or to the logical process step to which it belongs.

When navigating through multiple perspectives, it is important that the perspectives are linked in order to help the user track her current location in information space. Such a link can be a visual connection between two information items like the curves and arrows connecting process steps or an animated transition between two visualizations. These links should help the user to understand the semantic relation between the linked information elements. Figure 9 shows several items in two different perspectives: example ontology visualization and explosion view, connected by ThoughtFlashes (Einsfeld et al. 2006). ThoughtFlashes emerge from the item under the mouse pointer and rise in the form of 3D arcs to related items in an animated way.

One possibility to directly and intuitively connect process data to the process steps in which they are measured is the use of the landscape metaphor as shown in Fig. 10. The succession of the process steps is visualized as a planar graph (calculated with the Graphviz library). The height of the nodes is modified depending on the scalar value of the process parameter. On top of this height field, a landscape is calculated with Shepard's Scattered Data Interpolation method (Shepard 1968). As weight function we used the Euclidean distance. This method is often criticized

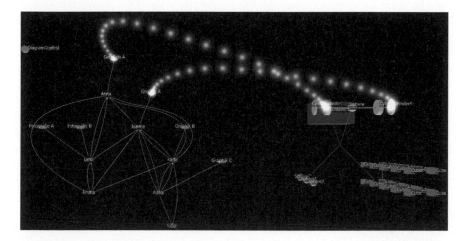

Fig. 9 Semantically interconnected perspectives

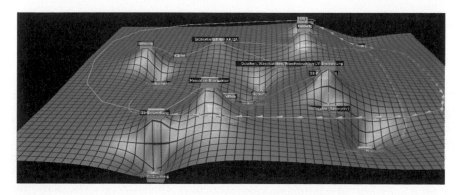

Fig. 10 Process data in landscape metaphor

for the fact that distant points have too much influence on the shape of the surface, interpolating local points. In our application, however, this is a volitional effect, as the mean value and the deviation from that value in each process step are easily perceptible. The progression of process parameters over time can be visualized with this metaphor as an animated deformation of the landscape. To give the user an additional hint at the actual value in a process step, we plan to use color as supplementary visual scale.

Interaction

Due to the fact that "getting lost in space" is a potential problem of 3D applications and that the mental load needed for navigating the viewpoint through the scene could be used for more important tasks, we proposed the ThoughtWizard navigation metaphor: the user relaxes at a fixed point in 3D spaces and uses simple gestures to move the needed information around him or her.

Our system supports navigation techniques like changing the visualization mode or perspective, scaling, rotating, or shaking the objects in the scene, or moving another process element to the focus position in ConeView. The interaction with the HANNAH framework is to some extent comparable to that of Blender, the 3D Content Creation Suite. The scene is directly manipulated (e.g., rotating or resizing objects or object groups) by mouse pointer gestures. There are several keyboard shortcuts that can be used to invoke certain actions or to change visualization modes. As an alternative, there are context menus displayed on demand directly in the scene. Thus, users can choose the interaction method that suits their needs.

As a matter of fact, direct user feedback is guaranteed in our dynamic 3D interface. Moreover, every transition is animated with a slow-in/slow-out movement as proposed by Thomas to reinforce the massive character of the visualization objects and to induce immersion.

Conclusions

SIMILAR has been evaluated on a convertible notebook (Toshiba Portegé 3500) and a Pocket PC (Compaq iPaq H5550). The Toshiba Portegé can be converted from a standard notebook to a Tablet PC. In the latter mode, all user input is done with a pen that is pointed onto the screen. The operating system (Windows XP Tablet PC Edition) features matured handwriting recognition, so no physical keyboard is required.

Our tests on the notebook showed that even users with little internet experience could easily access the system. The server response times are short and the bandwidth is sufficient for transmitting video sequences with a site up to about 10 MB. 3D performance for viewing VRML files was not convincing as the Toshiba

Portegé has only integrated a very simple graphics card (Trident CyberALADDiN-T with 16 MB UMA video memory). In Tablet PC mode, navigation is faster and more comfortable because hyperlinks can be directly activated by pointing to them with the pen. The handwriting recognition performs quite remarkably and is very reliable. Tests on the Pocket PC showed that the automatic adaptation to small screen sizes works very well. In most cases the relevant information can be viewed without scrolling or using special browser resizing features.

In addition to the technical approach of SIMILAR, KOMPLETT proposes the use of new information visualization metaphors that allow the intuitive understanding of complex information systems. The concepts of HANNAH are a modular architecture, flexibility, adaptability to changing needs, and intuitive access. The proposed toolbox of concepts and visualization metaphors can be used to develop interfaces to information rich environments with large amounts of heterogeneous but semantically related data. The animated 3D approach offers, besides more enjoyment and immersion in daily work, new possibilities like the animated 3D ExplosionView and the RotaryDiagram that are not available in today's information interfaces.

Initial feedback from people working with conventional systems reveals that they quickly recognize the potential of our visualization and are eager to evaluate it. Future work will be concerned with the development of more visualization and interaction metaphors, the integration of actual online-data from the plant, and first user evaluations.

Acknowledgements SIMILAR was supported by the *Stiftung Rheinland-Pfalz für Innovation*, Germany. The interdisciplinary project KOMPLETT is supported by the German Federal Ministry of Education and Research (BMBF).

References

Agrawala, Phan, Heiser, Haymaker, Klingner, Hanrahan, Tversky (2003) Designing effective step-by-step assembly instructions. ACM Trans Graph (22)

Ankerst, Kaim, Kriegel (1996) Circle segments: A technique for visually exploring large multidimensional data sets. Visualization '96, San Francisco

Ebert, Divivier, Barthel, Bender, Hagen (2001) Improving development and usage of visualization applications. IASTED International Conference on Visualization, Imaging and Image Processing (VIIP), Marbella, Spain

Einsfeld, Agne, Deller, Ebert, Klein, Reuschling (2006) Dynamic Visualization and navigation of Semantic Virtual Environments. IV '06: Proceedings of the Conference on Information Visualization. Washington

Goose, Güven, Zhang, Sudarsky, Navab (2003) Mobile 3D visualization and interaction in an industrial environment. Human Computer Interaction (HCI) International, Crete, Greece

Kincaid, Lam (2006) Line graph explorer: Scalable display of line graphs using Focus + Context AVI '06: Proceedings of the Working Conference on Advanced Visual Interfaces New York

Lipman (2003) Mobile 3D visualization for contstruction. International Symposium on Automation and Robotics in Construction, 19th (ISARC)

Schilling, Coors (2003) 3D maps on mobile devices. Workshop "Design kartenbasierter mobiler Dienste" Stuttgart, Germany

Shepard (1968) A two-dimensional interpolation function for irregularly-spaced data. Proceedings of the 23rd ACM national conference

Ware, Frank (1996) Evaluationg stereo and motion cues for visualizing information nets in three dimensions. ACM Transactions on Graphics (15)

Weber, Alexa, Müller (2001) Visualizating time-series on spirals. IEEE InfoVis Symposium

Simulation and Visualization of Indoor Acoustics Using Phonon Tracing

Eduard Deines, Frank Michel, Martin Hering-Bertram,
Jan Mohring, and Hans Hagen

Abstract We present an overview of our simulation and visualization system for sound propagation within closed rooms. For acoustic simulation we use a particle-tracing algorithm, the phonon tracing, which is suitable to simulate mid- and high-frequency sound. For low-frequency sound we have to fall back on FEM methods, which solve the wave equation. The result of the phonon-tracing algorithm, the phonon map, is used as a starting point for the visualization of sound propagation and exploration of the influence of different scene surfaces on the acoustic behavior of the room.

1 Introduction

Despite the obvious dissimilarity between our aural and visual senses, many techniques required for the visualization of photo-realistic images and for the auralization of acoustic environments are quite similar. Both applications can be served by geometric methods such as particle- and raytracing if we neglect a number of less important effects.

Human visual receptors can distinguish only three different basis colors (red/green/blue). But, what is the acoustic "color" of a material? It can be described by a frequency-dependent absorption function, which is smooth in most cases, and thus can be represented by few coefficients. Thus, it appears to be harder to "hear" the acoustic colors of objects rather than seeing their real colors, despite the fact that we can precisely distinguish between different frequencies. By means of the simulation of room acoustics, we want to predict the acoustic properties of a virtual model. For auralization, a pulse response filter needs to be assembled for each pair of source and listener positions. The convolution of this filter with an anechoic

E. Deines(✉)
Institute for Data Analysis and Visualization, University of California, Davis,
2144 Academic Surge, Davis, CA 95616, USA, E-mail: edeines@ucdavis.edu

G. Steinebach et al. (eds.), *Visualizing Sustainable Planning*,
© Springer-Verlag Berlin Heidelberg 2009

source signal provides the signal received at the listener positions. Hence, the pulse response filter must contain all reverberations (echos) of a unit pulse, including their frequency decompositions, due to different absorption coefficients.

For the room acoustic simulation we developed a method named phonon tracing, since it is based on particles. Our method computes the energy or pressure decomposition for each phonon sent out from a sound source and uses this in a second pass (phonon collection) to construct the response filters for different listeners.

From our visualization, the effect of different materials on the spectral energy/pressure distribution can be observed. The first few reflections already show whether certain frequency bands are rapidly absorbed. The absorbing materials can be identified and replaced in the virtual model, improving the overall acoustic quality of the simulated room. Furthermore, we represent the individual particles on their positions at the room surfaces colored according to their energy/pressure spectra. This method provides insight into how the room influences the spread sound waves. In order to observe the propagation of different wave fronts for the first few reflections, we visualize them as deformed surfaces. For the visualization of late reflections and a time dependent look at the phonon map, scattered data interpolation is used. For the representation of acoustic behavior at a certain listener position, we use colored spheres deformed according to the received sound.

The remainder of our paper is structured as follows: In section 2 we review related work. Section 3 describes our simulation approach, considering sound energy as well as sound pressure. Different visualization approaches are provided in section 4, followed by a brief conclusion.

2   Related Work

In the theory of acoustics there are two main approaches to simulating the propagation of sound. The first approach is based on wave equations that are numerically solved, for example, with the use of finite element methods (FEM). The simulation results are very accurate, but the complexity increases drastically with the highest frequency considered, since a volume grid with $O(n^3)$ cells needs to be constructed where n is proportional to the highest frequency. The time complexity for solving this is typically $O(n^3 log(n^3))$. Hence, the wave model is suitable for low frequencies only. The second approach, known as geometric acoustics, describes the sound propagation by sound particles moving along a directed ray. There exists a variety of such methods for simulating room acoustics. They are mostly based on optical fundamentals, and make use of approaches developed there. Two classical methods for acoustic simulation are the image-source method and the raytracing method.

The image-source method [1] models specular reflections by inserting additional sound sources obtained by mirroring the location of an audio source over polygonal surfaces inside the scene. The key idea here is that a sound ray from a sound source S reflected on a surface has the same acoustic effect as one from the mirrored source S', reduced by the absorbed energy part. The advantage of the image source method is, that it is very accurate, but it becomes very complicated for non-box-shaped rooms and curved surfaces cannot be simulated at all. The number of the mirrored sources increases exponentially with reflection order. Hence, this approach is suitable only for simple room geometry and low reflection orders.

In the raytracing method [10, 11] several rays are traced from the sound source to the receiver that is typically represented as a sphere. The reflections over the surfaces in the scene are simulated according to specular laws or Lambert's law of diffuse reflections. In addition to absorption properties of considered surfaces, timely delay and air absorption are taken into account. This approach is general and easy to implement, but it is very computation extensive and can cause aliasing artifacts due to discrete number of rays. When changing the receiver's position, the tracing of all rays needs to be performed again.

Due to the shortcomings of the two classical approaches described above, continuative approaches have been developed in recent years. Mostly, they employ parts of the classical schemes or a combination of them. One approach that makes use of the advantages of the image-source method and raytracing is introduced in [14]. Here the visibility check of the image-source algorithm is performed via raytracing. Beam-tracing methods [5, 6, 12] overcome the aliasing problem of classical raytracing by recursively tracing pyramidal beams, implying the need for highly complex geometric operations. Other approaches utilizing the photon mapping [7] also exist [8], but do not address the visualization of acoustical parameters. The phonon tracing approach [2, 3] precomputes particle traces and records the phonons direction and energy at reflecting surfaces for the collection phase. The particles contribute to the room impulse response at given listener positions.

The question often arises of how best to demonstrate an appropriate visual representation of the simulation results. Stettner et al. [13] visualizes room acoustic quality properties such as clarity or spatial impression by use of specific icons. Furthermore, they color the room boundaries according to the pressure level. Khoury et al. [9] represents the sound pressure levels inside the room by means of color maps. Additionally, the authors analyze the precedence effect (or law of the first wavefront) by using isosurfaces. Funkhouser et al. [5] use visualization of source points, receiver points, pyramidal beams, reverberation paths etc., in order to understand and evaluate their acoustic modeling method. The system also provides the presentation of acoustic measures like power and clarity. Existing commercial systems[1,2,3] provide some tools for visualizing the computed acoustic metrics throughout the audience.

---

[1]http://www.odeon.dk
[2]http://www.catt.se
[3]http://www.bose.com

3   Acoustic Simulation

For the simulation of the acoustics inside a room with given geometry and absorption functions at the surfaces we use the phonon tracing algorithm described in [2]. A recent improvement of this algorithm [3] uses Gaussian basis functions with approximate partition of unity to represent wave fronts. This is obtained by constructing the basis functions on the unit sphere around a sound source and dilating them with the traversed distance of the associated phonons. In contrast to the original approach where quadratic attenuation is used because of the spatial particle density, the improved phonon trace facilitates both linear and quadratic attenuation for the simulation of energy and pressure, respectively. This is obtained by scaling the basis functions according to the wave front propagation. Due to partition of unity, both simulated fields are continuous and smooth.

The improved phonon tracing is comprised of two steps: The *phonon* emission step calculates the particle traces and stores phonons on all reflecting surfaces in a *phonon map* and the *phonon collection* step calculates impulse response filters at given listener positions. As a result of the *emission step* we obtain the *phonon map*, which represents all reflections of an emitted wave front by a large set of particles (phonons) that can be considered as individual micro-sources in analogy to the image-source method. For each phonon in the phonon map the following information is stored:

- pressure or energy spectrum $p_{ph} : \Omega \rightarrow \mathbf{R}^+$
- the virtual source q_{ph} from which we can calculate the phonon's outgoing direction v_{ph} and the traversed distance d_{ph}
- the phonon's current position pt_{ph}
- the number of reflections r_{ph}
- and the material index m_{ph} at the current position

In order to compute the impulse response at a given listener position l_i, each phonon, which is visible from l_i, contributes to the individual frequency bands, as described in [2]. In the present work, we are not concerned with filter reconstruction, but rather with the energy and pressure amplitudes at a given listener position l_i. The contribution of a specific phonon to the listener position l_i is denoted in equation 1 (pressure) and equation 2 (energy).

$$p_{ph}(t,l_i) = \frac{\rho_{p,tot} P_0}{d_{ph}} w\left(\angle\left(v_{ph}, l_i - q_{ph}\right)\right) \times \delta\left(t - \frac{d_{ph}}{c}\right)$$

(1)

$$\rho_{p,tot} = \Pi \sqrt{1-\alpha_j}$$

$$p_{ph}(t,l_i) = \frac{\rho_{e,tot} e_0}{d_{ph}^2} w\left(\angle\left(v_{ph}, l_i - q_{ph}\right)\right) \times \delta\left(t - \frac{d_{ph}}{c}\right)$$

(2)

$$\rho_{e,tot} = \Pi\left(1-\alpha_j\right)$$

where p_0 and e_0 are reference pressure and energy at 1 m distance from the source respectively, α_j are the absorption coefficients along the phonon path, $\delta(t)$ is a discrete Dirac impulse shifted by the time elapsed between emission and reception of a phonon, and w is a Gaussian weighting function,

$$w(\phi) = \frac{2}{n_{ph}\sigma^2} e^{-\frac{\phi^2}{2\sigma^2}} \tag{3}$$

The radius σ can be chosen such that all Gaussian basis functions approximately sum up to one on the unit sphere. We note that there is a trade off between the smoothness of the partition of unity and the resolution of geometric shapes that can be represented in this basis.

4   Visualization Approaches

The phonon map characterizes the acoustic behavior of a scene considering the location of a specific sound source. It consists of the reverberations of a unit pulse, coming from different directions with different time delays and specific energy distributions. How can we visualize this complex information? In this section we briefly review different visualization approaches that are presented in more details in [2, 4].

4.1   Wave Front Propagation

Our first visualization method focuses on the spatial propagation of sound waves from the source. The corresponding wave front traverses the room and is reflected on surfaces, altering its intensity and energy spectrum. We visualize these sound waves by rendering small spheres representing the phonons. These are color coded by means of their spectral energy. Therefore, we use the RGB components, such that blue corresponds to the average of the energy by 40, 80, 160, and 320 Hz, green corresponds to the average of energy by 640, 1280, and 2560 Hz, and red to the average by 5120, 10240, and 20480 Hz. When sliding through time, the spheres follow the simulated phonon paths. We integrated the following functions in our interactive visualization system:

- varying the percentage of phonons to be rendered
- rendering only the phonons reflected from a selected material
- exchanging selected materials
- varying time / transversed distance

To improve the visualization of wave fronts, we use polygonal surfaces with phonons as vertices. Figure 1 shows an example of the described visualization approach.

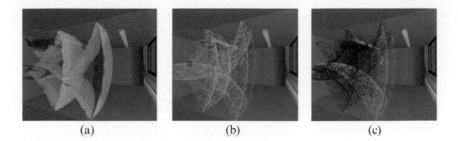

|(a)|(b)|(c)|

Fig. 1 Visualization of sound wave propagation. (**a**) Representation of wave fronts by use of triangulated surfaces. (**b**) Sound particles (phonons) rendered as small colored spheres. (**c**) Same as (**b**) the absorption properties of walls and ceiling are exchanged with absorption foam

4.2   Visualizing Phonons on Surfaces

The next method we have implemented to examine the phonon map is the visualization of certain phonons at their position inside the given scene. Each phonon is rendered as a sphere and is colored according to its spectral energy. We provide the option to consider all frequency bands in total or each of them separately. In the first case the spheres are color coded by using the RGB components, such as described in the previous section. In the second case, considering only one frequency band, we color coded the energy of this frequency band using the HSV model. We interpolate the color of the spheres between red (full energy) and blue (energy equals zero) corresponding to the energy $e_{p,i}$ of the i-th frequency band. In order to show the phonons outgoing direction v_p we render a cone whose peak is rotated towards v_p. The color of the cone corresponds to the phonons energy as well. Since the number of phonons in the phonon map is large, we render only phonons with a given number of reflections n_p simultaneously. Figure 2 shows an example of this visualization approach.

4.3   Reflected Wave Fronts

In this section we describe an approach visualizing wave fronts reflected at the room surfaces by use of triangulated surfaces (see Fig. 3.). In order to build these surfaces we need to know which phonons belong to a common wave front. Therefore we subdivide the phonons in clusters of equal history, such that phonons in the same cluster satisfy the following criteria:

- equal numbers of reflections n_p
- and for each reflection:
    - equal material indices m_p (same object of the scene)
    - equal surface normals n at the reflection position

Fig. 2 Visualization of
particular phonons

Fig. 3 Clustered wave fronts

Consequently all phonons inside a cluster have equal energy spectra. Depending on whether we want to examine all frequency bands in common or each of them separately, we use the RGB model or the HSV model as described in the previous section.

4.4 Scattered Data Interpolation

At higher reflection orders, the clusters become smaller and smaller, until they contain only a single phonon. In the following we develop a visualization method

Fig. 4 Energy on the floor at
20 msec for the 160 Hz band

for the entire phonon map based on scattered-data interpolation. The goal of the
interpolation is to get a continuous representation of the emitted energy on the
surfaces of the scene. First of all it is used to visualize phonons not belonging to
the early reflections of wave fronts, because these cannot be visualized as cluster
surfaces due to the increasing fragmentation. Rather than visualizing individual
reflections, we look at the phonon map from a different viewpoint, namely discrete
time steps, i.e., "show the energy emitted from the floor at 20 msec" (Fig. 4). This
allows us to look at the change in energy coming from a surface over time. The
interpolation is done for the energy and path length of the phonons. The direction is
neglected, since it is not important for the visualization due to scattering. The interpolation results are color coded corresponding to section 4.2 This means using
RGB color when visualizing the whole frequency spectrum and HSV for distinct
frequency bands.

4.5   Listener-Based Visualization

The approaches described above visualize the phonon map by considering the surfaces of the room and their acoustic properties. In this section we present a visualization method depicting the received energy at a listener position. With this
approach we can detect from which direction the most energy reaches the listener
and visualize the energy spectrum. For this purpose we render a triangulated sphere
deformed according to the weighted phonons received at the listener position. The
phonons are collected using the collection step described in section 3. For each
phonon that contributes to the total energy at the listener position, we first calculate
the intersection point $p_{intersec}$ between the ray from the sphere's center c_s to the phonon's position pt_{ph} and the sphere itself. We then increase the radial displacement
$disp_{sp}$ of that point sp of the intersected triangle whose distance to $p_{intersec}$ is minimal,
according to the energy contributed by the collected phonon. The color of the sphere
points is calculated according to the spectral energy distribution of the phonons
received at this position. Fig. 5. depicts the results of this visualization method.

Fig. 5 Deformed spheres representation for 1280 Hz band

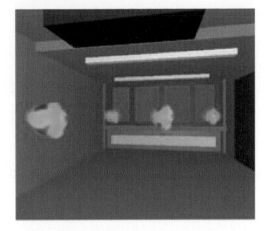

5   Conclusions

We have presented a particle tracing approach for acoustic simulation inside closed rooms as well as various visualization approaches for analyzing acoustic behavior by use of the resulting phonon map. We visualized the single phonons as color coded spheres. The advantages of this approach are the simplicity of the technique and the direct visibility of the material influence on the sound traversed from the source. A huge drawback is the lack of connectivity information between the phonons. This is overcome by the next approach, the visualization of wave fronts as triangulated surfaces. Additionally this method is a natural representation of the propagation of sound. Due to increasing fragmentation, this method can only be used for the first few reflections. To visualize the reverberations of higher order at the scene surfaces and to include the time dependency in the visualization, we used scattered data interpolation. At the moment, we restricted this method to the surface representation and neglect the direction of the particular phonons. In these methods we disregarded the situation at certain listener positions. Therefore we introduced the listener-based visualization approach. This technique allows a time-dependent view of the received energy at a certain listener position. In total, these approaches give a general idea of the acoustic behavior inside the considered scene that can be derived from the phonon map.

References

1. Allen JB, Berkeley A (1979) Image method for efficiently simulating small-room acoustics. J. Acoust. So. Amer 65(4):943–950.
2. Bertram M, Deines E, Mohring J, Jegorovs J, Hagen H (2005) Phonon tracing for auralization and visualization of sound. In IEEE Visualization. Minneapolis, MN.
3. Deines E, Bertram M, Mohring J, Jegorovs J, Michel F, Hagen H, Nielson GM (2006) Comparative visualization for wave-based and geometric acoustics. In: Transactions on visualization and computer graphics (TVCG), proceedings of IEEE visualization (12):1173–1180.

4. Deines E, Michel F, Bertram M, Hagen H, Nielson GM (2006) Visualizing the phonon map. In: Proceedings of EuroVis. Lisboa, pp 291–298.
5. Funkhouser TA, Carlbom I, Elko G, Pingali G, Sondhi M, West J (1998) A beam tracing approach to acoustic modeling for interactive virtual environments. In Computer graphics (SIGGRAPH 98). Orlando, FL., pp 21–32.
6. Funkhouser TA, Min P, Carlbom I (1999) Real-time acoustic modeling for distributed virtual environments. In: Computer graphics (SIGGRAPH 99). Los Angeles, pp 365–374.
7. Jensen HW (1996) Global illumination using photon maps. In: Rendering techniques '96 (Proceedings of the 7th Eurographics workshop on rendering), pp 21–30.
8. Kapralos B, Jenkin M, Millios E (2004) Sonel mapping: acoustic modeling utilizing an acoustic version of photon mapping. In: IEEE international workshop on haptics audio visual environments and their applications (HAVE 2004). Ottawa, Canada.
9. Khoury S, Freed A, Wessel D (1998) Volumetric visualization of acoustic fields in cnmat's sound spatialization theatre. In: IEEE visualization '98, pp 439–442 & 562.
10. Krockstadt U (1968) Calculating the acoustical room response by the use of a ray tracing technique. J. of Sound and Vibrations 8(18).
11. Kulowski U (1984) Algorithmic representation of the ray tracing technique. *Applied Acoustics* 18:449–469.
12. Monks M, Oh BM, Dorsey J (1996) Acoustic simulation and visualization using a new unified beam tracing and image source approach. In: Convention of the audio engineering society. ACM.
13. Stettner A, Greenberg D (1989) Computer graphics visualization for acoustic simulation. In: International conference on computer graphics and interactive techniques, pp 195–206.
14. Vorländer M (1989) Simulation of the transient and steady-state sound propagation in rooms using a new combined ray-tracing/image-source algorithm. *J. Acoust. So. Amer.* 86(1): 172–178.

III
Case Study

Digital Phoenix Project: A Multidimensional Journey through Time

Subhrajit Guhathakurta, Yoshi Kobayashi, Mookesh Patel,
Janet Holston, Tim Lant, John Crittenden, Ke Li, Goran Konjevod,
and Karthikeya Date

Abstract The Digital Phoenix Project is a multiyear project aimed at developing
a realistic digital representation of the Phoenix metropolitan area through space
and time that can be experienced in Arizona State University's Decision Theater.
A significant objective of this project is to create an environment for querying,
researching, and visualizing critical urban sustainability issues confronting a rapidly urbanizing area. By creating a multidimensional virtual model of Phoenix from
a variety of data sources, we can visualize patterns of growth and development, as
well as their consequences, emerging across the continuums of space and time. The
modeling of future environments will enable the assessment of policy scenarios and
guide desired future urban-environmental patterns. A digitally constructed model
of the city will also allow us to discover what Phoenix could have been like, starting
with historical data as the basis for projecting to the present and into the future.

1 The Genesis of Digital Phoenix

The Digital Phoenix Project started in July 2006 with a grand vision of developing
a digital representation of critical elements driving the evolution of the Phoenix
metropolitan region. By digitally encoding important information about the past,
present, and potential future of the metropolitan region, we hope to create a well-
integrated experimental model that can be queried and visualized in a highly
immersive environment like the Decision Theater at Arizona State University. The
objective is to generate well-calibrated and realistic experimental models that can
provide answers to questions such as: a) why and how the city developed in the
manner it did, b) how the decisions we make today will shape future environments

S. Guhathakurta(✉)
School of Planning, College of Design, Arizona State University, Tempe, AZ 85287-2005, USA

G. Steinebach et al. (eds.), *Visualizing Sustainable Planning*,   159
© Springer-Verlag Berlin Heidelberg 2009

in this region; and c) what will be the sustainability implications of the future scenarios. In essence, we are attempting to build a tool for planning and decision-making that can be used by public officials, scholars, and citizens, to enable sustainable growth in this region.

We began the project by parsing the broad agenda into three component parts: 1) a digital 3-D parametric model of the current physical environment; 2) digital encoding of historical data on the physical and environmental evolution of metro Phoenix; and 3) an urban futures simulation engine that can provide scenarios of future land use, housing, travel, and employment patterns based on current trends and policy choices. While the three components listed above continue to be the primary branches of the Digital Phoenix tree, several other research efforts were also brought into the fold to resolve more fundamental issues tied to the objectives of the project. For example, we required spatial and non-spatial measures of sustainability to compare the various current and future developments (scenarios) in Phoenix. Also, the voluminous amounts of data and the corresponding high intensity of computation that is required necessitate high performance computing. We were fortunate to have the assistance of ASU's High Performance Computing group who helped in leveraging our computing power severalfold. In addition, a team of individuals at the Decision Theater has been critical to this endeavor given the ultimate objective of porting many of our products to the immersive environment offered by the Decision Theater. The integration of the various components of the Digital Phoenix Project is illustrated in Fig. 1.

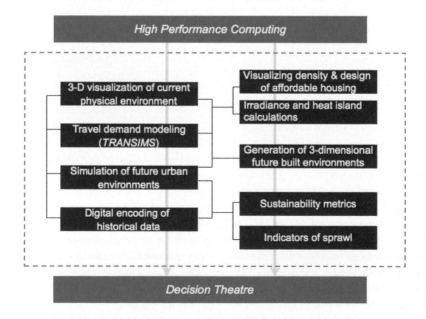

Fig. 1 Integration of the different components of Digital Phoenix Project

2   The Context

Human civilizations in central Arizona date back over 2000 years, although the current history of Phoenix starts sometime in the nineteenth century. The older civilizations were organized around water and other natural resources that were necessary for their survival. The modern city of Phoenix is also dependent on water and resources, although massive infrastructure and urban planning processes are necessary to support the population as the city grows. With growth comes increasing complexity and challenging decision-making regarding future urbanization. Economic development, a creative workforce, and technological innovation are critical assets in continuing the transformation of Phoenix into a more desirable city. The Digital Phoenix Project will create a visual planning tool with keen insight into urban dynamics in Phoenix through the use of state-of-the-art visualization, computation, and informatics tools, combined with detailed social, economic, and environmental data.

Rapid urbanization of the Phoenix metropolitan area has led to increasing pressures on the area's environment and infrastructure. Greater Phoenix is in the grip of explosive growth. The region's home county, Maricopa, sits atop the list in U.S. population growth from 1990 to 2000. The Phoenix-Mesa Metropolitan Statistical Area (MSA) ranked 5th in terms of absolute population growth in the U.S. over the same period, and 8th in percentage growth. The population build-up has even taken hold of smaller cities. Gilbert ranked 2nd in population increase across all U.S. incorporated places, with Chandler (9th) and Scottsdale (15th) following close behind (see Fig. 2 for location map). According to unofficial projections, Phoenix is currently the fifth largest metropolitan region in the United States having surpassed Philadelphia sometime in 2005 (Wikipedia). Several of these cities, such as Paradise Valley and Sun City, are relatively small enclaves; others such as Phoenix, Mesa, and Scottsdale are large conurbations. Given the rate of growth and the diversity of settlements in terms of their history and socio-economic characteristics, the region is an archetype for studying the sustainability debate facing urban systems across North America. Indeed, Phoenix imports over one million acre-feet of water per year to support the population and its agricultural and industrial activities.

The metropolitan area consisting of Maricopa and Pinal counties spans a combined land area of 14,598 square miles for about 3.5 million inhabitants with a resulting population density of 184 persons/sq-mile. Like most cities, the metropolitan region includes a textured surface of human demographics with regions of more and less affluence, age-distribution, population density, and occupational communities including many ethnic communities. These unique features make Phoenix an important city to explore because, in many ways, it is the first desert city to grow so quickly for so long to such a tremendous size. No other city in the U.S. is hotter, making Phoenix a prototype for studying rapid urbanization in desert regions, a trend that is occurring globally in other desert locales. Due to the rapid growth, the planning, development and management, of infrastructure in Phoenix is complicated. Features of the city are changing so rapidly that unless a thorough documentation of

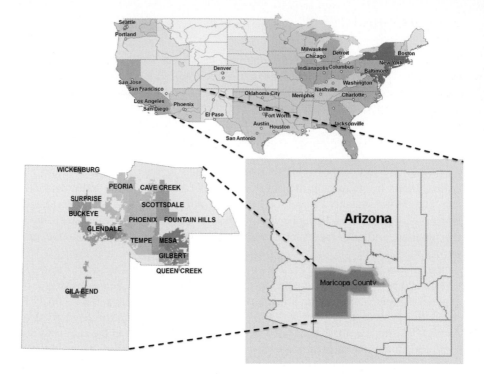

Fig. 2 Location map of cities in Maricopa County, Arizona

the data-sources are captured during this transformation we will never have the opportunity to understand the successes and difficulties involved with building this metropolitan region.

The pressures of this explosive growth in the Phoenix metropolitan region are evident across various urban systems including transportation, water distribution, and temperature controls. The demands on urban infrastructure are growing disproportionately faster than the rate at which such infrastructure is being expanded. The environmental problems of air pollution, water availability, and land degradation, among others, are compounded by both the increasing demands on natural resources by the growing urban population and the inability to manage such population growth through the adequate provision of urban infrastructure (Guhathakurta 2003). What is needed is a new approach to urban planning, design, and environmental management. This approach should use the current methods and techniques, but also include modern technology and state-of-the-art science to create realistic and compelling models of urban features that can be explored interactively by planners and key decision-makers. Such tools will fundamentally change the way cities are designed and managed. This approach will also enable innovative thinking and rapid exploration of the connections between land, the environment, human behavior, and human settlements. It will also necessitate new tools and better theories to understand the dynamics of urban growth and change. These are the goals of the Digital Phoenix Project.

The development of visualization tools has several potential uses. For example, it would allow us to examine the factors that led to the city's development patterns based on historical information. The multilayered historical digital data will provide answers to several questions regarding "evolutionary trends" such as changes in land resulting from the loss of streetcars or from the advent of new technologies (e.g., air conditioning). Specific questions about influences on landscape patterns, household water use, and changing demands for energy and materials can also be examined using tools developed for Digital Phoenix. Further, simulation models of future growth in Phoenix can also offer important insights about desirable social behavior and policy choices. The Phoenix UrbanSim model will be able to evaluate questions such as: how will the future light rail transit system in the Phoenix metropolitan area affect the type and location of new housing (a topic that will be discussed later in this article)? Where will the concentrations of jobs be located if Phoenix becomes a hub for biotechnology companies? What will the air quality in the metropolitan area be if automobile usage continues to increase at current rates? The answers to many of these questions require detailed information that is currently becoming available and will be brought together under this project. Most importantly, as this project matures, we will create a platform to share possible outcomes with key decision-makers, researchers, and the general public to support informed decision making about the future of Phoenix.

Throughout its history, Phoenix has undergone significant transformation in its structure and functions, and in its social, economic, and physical environment. This transformation, which up until now has been continuous in some respects and discrete in others, is impossible to predict from any point in the past. More recent research on cities as self-organizing, complex systems reinforces the fact that the future is not given and deterministic laws do not, by and large, govern the evolution of human habitation on this planet (Batty 2005; Beneson and Torrens 2004; Portugali 2000; Guhathakurta 2002). What we do know is that massive modern cities create planning challenges that require foresight into the possible consequences of current decisions. Cities are now being examined from the perspective of networked systems stretching in time and space. These networks evolve in dynamic, self-organizing ways, generating structure while at the same time breaking symmetries. We now know that there is no universal optimization principle for complex systems, like cities, but several possible futures that differ from each other qualitatively.

The core of the Digital Phoenix vision is a digital copy of the Phoenix metropolitan region in space and time that can be experienced in the Decision Theater and other richly textured immersive environments. Elements of this vision include a visualization component that would scale according to the users perspective and include large libraries of richly textured urban elements that can be combined to construct both local and regional urban environments. These environments would be amenable to visual exploration at different scales and perspectives with appropriate walk, drive, and fly-through digital techniques. The second component of

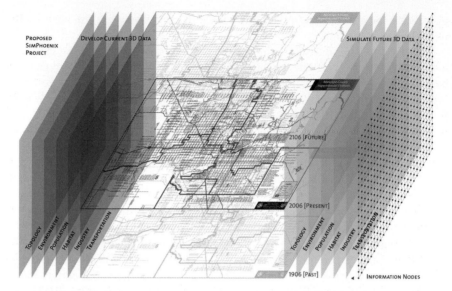

Fig. 3 Conceptualizing Digital Phoenix as a "journey through time"

the larger vision is to develop an engine to model and visualize both the past and the future using detailed data from historical and current social, physical, economic, and environmental attributes. Different techniques are being used to project the future and the past. While future projections will use statistically calibrated, agent-based models of individual and social behavior across space and time, the reconstruction of past environments will be based on detailed archival analysis of several "layers" of data (see Fig. 3). By creating a multidimensional representation of urban data, we will be able to visualize patterns that emerge in space and time. That is, one goal of the visualizations is to create realistic 3-D images of the city for navigation purposes, while other visualizations will reveal patterns of economic development through time. The ultimate longer-term objective is to bring together the modeling and richly textured visualization components so that we will be able to visualize within one environment, the past, the present, and the future. Hence we can recreate the evolution of Phoenix as a journey through time.

3   Research Components and Methods

The project has three main components: the development of a digital representation of Phoenix in the present, historical records of Phoenix that can recreate the city at times in the past, and modeling tools and theories to generate future representations of Phoenix. All three components are ambitious in scope, but by integrating the three into a unified framework and leveraging the Decision Theater capabilities, the outcome promises to be a powerful tool for urban planning in Phoenix.

As we create the ability to digitally-document the present, we will also begin to use the information about the present to extend into the past and future. Historic records exist, but not in digital data formats because the technology was obviously not available. With the advent of computers, historical information is becoming more available in digital formats, but often important qualitative information and details that do not easily convert to digital formats are lost. This project aims to advance the digital archives of Phoenix's history and use the capabilities of visualization to move back in time and track its progression. We will create a digital past of Phoenix in such a way that the data is integrated with future information. We will also create a digital future from the data and the Phoenix UrbanSim implementation. The ability to move forward in time from current information will also generate the capability to move forward in time from past states. This capability will allow us to fill in the periods of time when limited records were available though theories of settlement and urbanization. It will also allow us to envision what the future might have been like if different decisions were made in the past.

The project has several components but the central core is comprised of a rich database that scales up and down both in time and space. This data is then selectively extracted to develop multidimensional visualization of large regions as well as small neighborhoods. The data would also provide information about the evolution of various systems of social, physical, economic, and environmental networks over time in the Phoenix metropolitan area. This would allow the examination of the history of specific social habitation patterns, offer means of experiencing real time-immersive environments, and provide a means for evaluating future growth scenarios. Each of these endeavors would be developed concurrently in a manner such that they integrate across scales and times. The three components of this project are discussed below.

3.1 Visualization Through Immersive Environments in Real Time

One of the most important aspects of visualizing scientific simulations of a city is the definition of 3-D data structure of the virtual city model. To date, no data format allows for a comprehensive picture of all urban processes. Nor do we expect that such a representation of reality can exist. However, the tools and technologies that we work with as part of our current research will allow for significant advances in capturing many of the fundamental processes that emerge in complex urban systems. Moreover, the creation of appropriate digital representation has tremendous value for pattern recognition, and for rapid theory testing and design, on a synthetic population whose basic actions are well-predicted at large scales. Having the basic data structures and initial set of real data is an incredibly powerful capability, but it is only the beginning of what is possible in large-scale urban simulation. Especially, in order to visualize the model for decision-making in immersive VR (virtual reality) environments, the data structure should be designed to support various kinds of input data flexibly. For example, some decision makers may need to observe voting patterns from the current city model. Others may need to experiment with new buildings for landscape design evaluation.

In addition to the data structure issue, the city model must be visualized with enough impact for decision-makers to explore new ideas and help deepen observation. Therefore, the city model should be created with textures and details.

There are many different methods for creating 3-D city models, and researchers are trying to develop more efficient and effective methods. These modeling methods are mainly categorized into three approaches; automatic, semi-automatic, and manual. The automatic approach is to extract 3-D objects such as buildings, streets, and trees from aerial or satellite images by using the technologies of image processing and pattern recognition in artificial intelligence. The semi-automatic approach is to create 3-D objects one-by-one with the support of technologies like photogrammetry and 3-D vision. The manual approach is to create all geometries of an object one-by-one in Computer Aided Design (CAD) and Computer Graphics (CG) packages that are commercially available such as 3-D Studio Max and Maya. Spine3D is one of the well-known CG design platforms that provide a means for developing 3-D city models manually.

The methods of 3-D city modeling also vary according to the resources available and overall objectives of the project. LiDAR (light detection and ranging) and photogrammetry are the technologies commonly used in extracting 3-D geometries. The LiDAR instrument transmits light to a target and measures its dimensions by using the reflected signals. There are two approaches in using LiDAR. One is to acquire the LiDAR data from an airplane. This is commonly used in remote sensing for creating digital surface model (DSM) and digital terrain model (DTM). Another approach is to get the LiDAR data from the ground and extract the complicated geometries like architectural components and civil structure (Früh and Zakhor 2003). A set of points extracted from LiDAR is converted into polygons. This procedure makes it possible to obtain details, but it requires researchers to fly or walk over to get the necessary data.

Photogrammetry is another solution for extracting 3-D geometries. Like LiDAR, it can be used for both aerial and ground images. Aerial images are used to extract abstract forms of buildings, and the ground images are used to extract their details. Nverse Photo (www.precisionlightworks.com) and Shape Capture (www.shapecapture.com) are examples of commercial software packages for 3-D modeling that use photogrammetry. There are two approaches for photogrammetry. One is to use photos taken from the ground. An advantage of this approach is the relative low cost. On the other hand, there are some disadvantages. First, it is difficult to take photos of buildings from behind because of security and privacy issues. Second, since several photos are required to cover all elevations for each building, it is necessary to manage a number of image files. Third, it is difficult to match the images with different white balances on the same building. In short, the approach using

Table 1 Classes of 3-D City Models

	Low Quality (Online Quality)	Middle Quality (PC Quality)	High Quality (Movie Quality)
Street Level	SL	SM	SH
Block Level	BL	BM	BH
City Level	CL	CM	CH

ground photos is useful for extracting building geometries when details are required such as for SH and CH models (see Table 1), but this process requires more time to manage and fix the textures.

Yet another approach is to use aerial or satellite images. This process is sometimes advantageous because it requires only a few images. Since the textures of buildings are extracted from the same image, the color balances in the image are not an issue. However, a substantial disadvantage is the cost of this method; it costs more to take aerial photos than ground photos.

3.1.1 Classification of 3-D City Models

Choosing the most suitable method for creating 3-D city models depends on the given resources and objectives. Table 1 illustrates one classification of 3-D city models based on quality and spatial scale.

There are three scale categories: street level, block level, and city level. The street level model is used to visualize a street with buildings and landmarks (such as trees, traffic lights, signs, and bus stops) from a human perspective. The block level model visualizes street blocks in a city, including buildings and landmarks, from a bird's eye view. The city level model visualizes a whole city from several thousand feet above ground as from an airplane.

In addition to the classification based on scale, the 3-D city models are classified into three quality classes: low, middle, and high. The low quality model is designed to render interactively in real time on Internet browsers, the middle quality model is to render in real time on PCs, and the high quality model is not for interactive rendering but for static rendering.

- The low quality street level model (SL) has buildings and landmark components without any textures or materials. This model is typically used for evaluating the height and volume of buildings from a human perspective. This method is often applied at the beginning phase of design in architectural design studios. The model is created with commercial 3-D computer graphics software packages such as FormZ (www.formz.com) and SketchUp (www.sketchup.com).

- The middle quality street level model (SM) has more details and textures than SL. Many 3-D games such as DOOM3 (www.doom3.com) is classified in this category. In order to visualize the model interactively in real time, the details are created with minimum polygons. With the improvement in graphics card technologies, very realistic images can be rendered with high-resolution textures.

- The high quality street level model (SH) is the highest quality model and is seen in architectural presentations and Hollywood movies. Since it is necessary to create 3-D objects one-by-one using computer graphics (CG) packages, it takes a lot of time and effort. The images are very realistic and beautiful, but it cannot be rendered in real time.
- The low quality block level model (BL) is used for visualizing street blocks in a city. Since a model usually has many buildings, each building is a simple volume without any textures in order to render them in real time. Google earth is one example that shows this model in 3-D views (http://earth.google.com/). 3-D-GIS models with digital terrain model (DTM) and 3-D buildings, which are created by extruding 2D polygons with building height values, is classified in this category as well.
- The middle quality block level model (BM) is an upgraded form of BL with textures for buildings and ground. The ground object has textures based on ortho-images. Many automatic approaches have been researched and developed for creating models in this class using photogrammetry and image processing technologies (Pennington and Hochart 2004).
- The high quality block level model (BH) is based on BM, but with more details added to each building. The model is usually employed for static rendering because it is too computing-intensive to render the model interactively in real time. A model of 1930s New York City used in the Hollywood movie *King Kong* is an example of this class.
- The low quality city level model (CL) shows only DTM mapped with ortho-image without buildings, street, or landmarks.
- The middle quality city level model (CM) has DTM and buildings without textures. Each building is represented as a box.
- The high quality city level model (CH) has DTM and buildings with textures. These are extremely difficult to render in real time without scaling (i. e., changing the quality of resolution according to scale).

3.1.2   Creating 3-D Models of Downtown Phoenix

In this project, we adopted the method of creating 3-D city models from aerial photos using photogrammetry. This approach is chosen because it allows the development of a model that can be used in eight classes described in Table 1 (SL, SM, BL, BM, BH, CL, CM, and CH). This approach is also less time-consuming than the other approaches discussed.

The processes of taking aerial photos, scanning images, extracting buildings from the images, and editing 3-D objects are explained below.

Aerial photo and scanning. The most important step in creating 3-D city models from aerial photos is to acquire high quality images in order to extract the textures of buildings.

In this project, several different kinds of aerial photography techniques were investigated. Two different cameras were tested. One was Canon Eos-1Ds Mark-II, which is capable of taking the highest resolution image among the digital cameras commercially available. The images were shot from an altitude of 6000 ft by airplane and from 2000 ft by helicopter. During the flight, an aerial photographer held the camera and took images manually.

In addition to the Canon Eos-1Ds Mark-II, a regular aerial photo camera for 9" × 9" negative films was used to take the photos from 6000 ft and from 10,000 ft. The camera was fixed on the airplane and a pilot released the shutter.

As explained in Table 2 below, the image taken from 6000 ft using the regular aerial photo camera was the best for this project. The other images did not provide adequate details because the side images of buildings were not clear enough to detect building textures.

In order to get clearer images of the textures of the sides of buildings, it was necessary to take several oblique shots in addition to the vertical shots. The flight path for taking regular stereo-pair aerial photos is usually straight, as shown in the left image of Fig. 4. However, in order to take several oblique shots of the same target, the airplane needs to fly over the same position repeatedly as shown in the right image of Fig. 4. In addition, the pilot needs to release the shutter for each shot looking at the screen monitor in order to check the position of target in the image, and the photos are taken during a circular flight. The pilot needs to be skilled in order to get the proper oblique aerial photos. Three flights were required for this project since the first two failed because of an inexperienced pilot.

Table 2. Choosing appropriate aerial photographs

	6000 ft Film	10,000 ft Film	6000 ft Digital	2000ft Digital (Helicopter)
Pros	Best quality	Cover 3 × 3 miles	Easy shot	Easy shot
Cons	Need an expert pilot	Low quality	Blurry images	Blurry images Many images

Fig. 4 Flight path for stereo shots

Fig. 5 Aerial images scanned at 2000 dpi

Each aerial photo is scanned at 2000 dpi resolution, and saved as TIF formatted image without any compression. Each scanned image has about 18,000 × 18,000 pixels as shown in Fig. 5.

Modeling process. Nverse-Photo 2.7 (www.precisionlightworks.com), one of the commercial off-the-shelf photogrammetry tools, is used to extract 3-D buildings and the ground from aerial images. First, the following camera parameters are defined for stereo matching: 1) calibration focal length = 152.884 mm, 2) lens distortion is input as

$$K_0 = -0.2877 \times 10^{-6}, K_1 = -0.8168 \times 10^{-8}, K_2 = -0.4265 \times 10^{-22}, K_3, K_4 = 0.0000$$

3) the scanning resolution is set at 2000 dpi; and 4) X and Y offsets are defined using fiducial marks.

Once camera registration is done for all aerial images, the stereo matching process progresses smoothly. After the matching process, each building is created by drawing polygons on all images. For example, a polygon is created in the first image, and it is sent to the second image. The user needs to edit the polygon in the second image corresponding to the building. This process defines the height and form of the building. By repeating this process, the geometries of buildings are defined, and the textures of building are automatically assigned from different parts of aerial images. The ground is defined by inputting the ground truth points with the information about latitude and longitude instead of polygons. The texture for ground is also generated automatically.

The model is saved as a 3DS formatted file, which is one of the most common 3-D formats. In order to visualize the model at 3-D stereo theater, the model is converted from 3DS format to OSG (http://www.openscenegraph.org/) format. Fig. 6 shows the workflow of this project.

Editing 3-D model. Autodesk 3D Studio Max is a professional 3-D computer graphics modeling and rendering package. The package is used for editing the buildings when they have some problems regarding their textures. If some parts of buildings are not visible in some images, the textures are distorted or not generated.

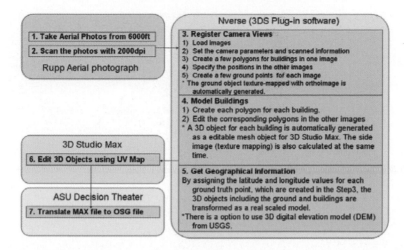

Fig. 6 The Modeling Process

Fig. 7 Editing buildings

In such cases the images taken from the ground can be used to fix the problems. UV-mapping, which is one of the most advanced techniques commonly used in developing 3-D games, is applied to edit the side images of buildings as shown in Fig. 7. This process is tedious because it takes a few hours to fix the problems on each building.

3.1.3 Results and Application

Fig. 8 shows the model created in this project. The model covers about one square mile (1.6 km × 1.6 km) area in the center of downtown Phoenix with more than 700 buildings. The model was created within 16 hours by one person.

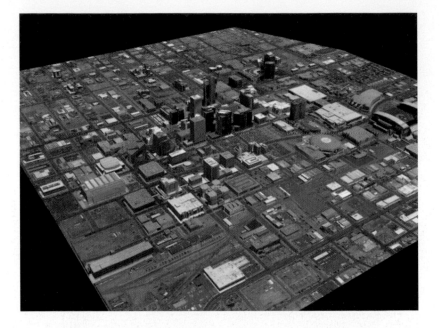

Fig. 8 3-D city model of downtown Phoenix

Application of 3-D city **model in Digital Phoenix Project.** The Digital Phoenix Project is developing a 3-D immersive model of Phoenix's built environment on a UC-Win/Road platform (http://www.forum8.co.nz/index.php/eng/home). The immersive environment is specifically designed for Arizona State University's Decision Theater. This model currently comprises the one-mile square section of the downtown area (shown in Fig. 8), which has been detailed with appropriate textures, street furniture, and signs, together with drive-through capability (see Fig. 9). The transportation infrastructure of Maricopa County is now being imported to the platform for tying together sections of the built environment that have already been constructed in UC-Win/road or are in that process. The transportation infrastructure will also articulate with TransSims, traffic simulation package to provide accurate traffic counts at specific time of day during the week. By populating the virtual downtown environment with appropriate traffic counts, we will be able to offer realistic experiences of navigating through traffic at various times during the day. This will offer excellent first-hand knowledge of the subjective experiences of drivers and pedestrians in the Phoenix downtown area.

Other Applications of 3-D models. Four other applications that used the 3-D city model of downtown Phoenix are shown in Fig. 10. The top left image (10a) shows the online web 3-D application using this model (www.ruthron.com/purl). The user can change the views and get the information of buildings on Internet browsers. The bottom left image (10b) demonstrates the application of 3-D printing. By using the device to get the XYZ position on the physical model, the building information is displayed. The top right im age (10c) is the conceptual image

Fig. 9 A scene of Phoenix downtown from UC-Win/Road

Fig. 10 Application of 3-D city mode: 10a (*top left*) online application; 10b (*bottom left*) tangible interface; 10c (*top right*) physical model; 10d (*bottom right*) VR model

to integrate the applications 10a and 10b with a big 1/32" scaled physical model. The bottom right image (10d) depicts a viewing of the 3-D city model in a VR environment.

3.2   Modeling of Phoenix Urban Futures

An important component of the Digital Phoenix Project is the ability to model and visualize alternative futures of Phoenix-based on different choices made about policies, regulations, and lifestyles. The simulation of future developments and their implications for environmental conditions and resource availability offers critical insights into how the future will unfold based on choices made today. Over the past several years we have implemented a modeling environment called UrbanSim, which is under active development at the University of Washington at Seattle. UrbanSim is probably the only large-scale agent-based model that has been successfully tested and implemented by regional and academic communities, which simulates future urban growth at a fine-grained spatial scale. We currently have a model of Maricopa County at a spatial resolution of one mile (Joshi et al. 2006). We are also implementing another version of the model at a spatial resolution of 150 meters.

3.2.1   Policy Analysis Using Urbansim

The Central Phoenix/East Valley Light Rail Transit Project, which is now under construction, will provide convenient and comfortable transportation between Phoenix's central business district, the Sky Harbor International Airport, Arizona State University, several community college campuses, and event venues that currently draw about 12 million people each year from the region. The first phase of the project will include a 20.3-mile line that connects significant destinations in three cities – Phoenix, Tempe, and Mesa. It is expected that this phase of the project will be completed by 2008. In light of the new transportation option that will become a reality in less than three years, planners in the three cities are actively engaged in planning and redesigning the areas around the transit stops. The scenarios tested in this study takes into account many of the planned interventions around the transit stations, mostly in terms of introducing mixed-use and higher density developments. Figure 11 shows a map of the overall planned system in relation to the various cities in the Phoenix metropolitan area.

The first phase of the Phoenix Light Rail project will include 32 transit stations within the cities of Phoenix, Tempe, and Mesa. These station areas are shown in Fig. 12. Given that transit stops are designed to be closer together than the one-mile grid used in the UrbanSim model, we have allocated three analysis zones each having distinct characteristics. Zone 1 radiates north from downtown Phoenix and

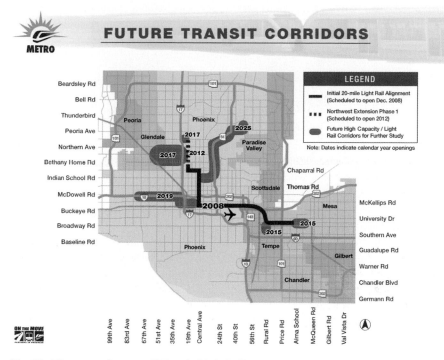

Fig. 11 Alignment of proposed Phoenix Light Rail

includes most of the downtown business district and the uptown arts district. This region includes some of the oldest neighborhoods in this metropolitan region and a fairly large downtown core. Zone 2 is currently a low-density corridor that is adjacent to the commercial airport and includes many industries that have located to take advantage of proximity to the airport. This corridor also includes several low-income neighborhoods and areas with a high concentration of minorities. Zone 3 is dominated by Arizona State University and activities supporting the university clientele. The concentration of student housing is high in this area. This zone also includes several ethnic retail establishments catering to a large international student community attending Arizona State University. The following analysis compares the transition of households in the three delineated zones based on scenarios with and without light rail transit for year 2015.

The scenario for different levels of transit usage was generated by changing modal split for all the transport analysis zones (TAZs) that include the 32 stations mentioned earlier. Accessibilities were recalculated such that for the 5 percent scenario, 5 percent of the total number of trips was added to transit and subtracted from auto. Similar procedure of increasing transit ridership was adapted for 15 percent and 25 percent scenarios. These scenarios were tested against "no build," where light rail is not built and the existing mode split continues into the future.

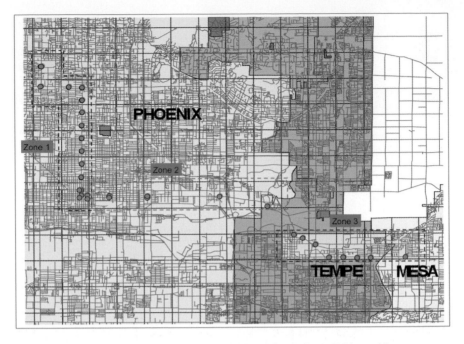

Fig. 12 Delineation of zones for styudy of light rail impacts on household transition

Also, cities will be rezoning the station areas for high density, mixed-use developments. To account for this land-use change, development types of the gridcells falling under stations have been changed to high density and mixed-use development type. The particular light rail scenario discussed below assumes the mid-range of the three scenarios tested, that is, 15 percent of trips to and from the areas adjacent to light rail will be on the proposed Phoenix Light Rail system.

Implementing UrbanSim: data, process, and validation. UrbanSim is not a single model. It is an urban simulation system, which consists of a family of models interacting with each other, not directly, but through a common database. There are seven different models within UrbanSim (economic transition model, demographic transition model, employment and household mobility models, employment and household location choice models, household mobility model and the real estate development model). A more detailed description of each of these models and their underlying theories are available at (http://www.urbansim.org/index.shtml). Here we discuss the specific implementation of UrbanSim for Maricopa County, completed at a spatial resolution of one mile, and its application in evaluating one scenario of future land use and household type changes resulting from a proposed light rail system.

Data.
The input data included in the data store consists of parcel file information from the county assessor's office, employment data from Maricopa Association of

Governments (MAG), census data, detailed land use and land valuation data, boundary layers showing environmental, political and planning boundaries (also from MAG) and a chronological list of development events. A set of software tools such as ArcGIS and MySQL was used to extract the data from input files, calculate values and construct the model database in the specified format.

The data store contains all households in Maricopa County starting with the base year 1990. Each household is a separate entry in the households table with associated characteristics such as household income, size, age of head of household, presence and number of children, number of workers, and the number of cars. In addition, the data store contains every job present in the Maricopa County by location (i.e., grid id), job sector, and whether the job is home-based or not. Altogether UrbanSim requires about 60 data tables, which are used in the complete database. Each table has a well-defined structure. The model components include a script to check consistency of tables, which when run provides warnings and error messages if any table is incorrectly formatted for UrbanSim.

The process of creating the Maricopa County database for UrbanSim required the following steps:

1. Define project boundary (Maricopa County)
2. Define the base year for data (we used 1990)
3. Generate grid (9511 grid cells, one-mile by one-mile each, were generated for Maricopa County)
4. Assign unique IDs to grid
5. GIS Overlays: Parcels on grid; transportation analysis zones (TAZ) on grid
6. Allocate parcel characteristics to grid
7. Assign employment to grid
8. Reconcile non-residential space and jobs
9. Synthesize households and locate them by Grid ID
10. Generate diagnostics and resolve inconsistencies
11. Assign development types
12. Convert environmental features to grid
13. Convert planning boundaries to grid
14. Load database into MySQL
15. Run consistency checker

In this paper, we do not attempt to describe each of these steps in detail but the steps are well documented in the UrbanSim manual available at www.urbansim. org. There is however one step, step 9 above, that requires special attention given that it is an extraordinary process when compared to most land-use projection models. As mentioned earlier, UrbanSim database requires information for every household in Maricopa County by their special attributes. There is no single source from which this entire data can be obtained. For this reason, UrbanSim provides a utility called the Household Synthesis Utility. As its name suggests this utility synthesizes the household data with the help of an *iterative proportional fitting algorithm*. The utility synthesizes households separately by family type for each Public Use Micro Area (PUMA) at the level of the block groups. Data sources required for the household synthesis utility are: 1) Sample of households by age of head, income, race, workers, number of children, and number of cars for families as well as non-families at the level of PUMAs from 5 percent Public-Use Micro-data; and 2) block group level data for the marginal distribution tables from U.S. Census Summary Tape file STF-3A. The algorithm iteratively matches the marginal totals at the level of block groups to varying sets of households represented in the 5 percent PUMS sample. When a selected set of households match closely the aggregate block group statistics, that household set is assumed to belong in that block group. Subsequently, households in the block groups are associated with the grids-IDs. In this manner the households table is populated and the synthetic process of household allocation closely approximates actual household locations.

Another important aspect of this modeling approach is the use of accessibilities as a critical driver of jobs and household locations. The information about trips between TAZs in Maricopa County at various points in time was obtained from Maricopa Association of Governments (MAG). This data allowed us to calculate logsums by travel mode from which accessibilities were derived for incorporation into UrbanSim data store. UrbanSim is usually run in tandem with an external travel model, so that the accessibilities can be updated at regular intervals. For our purposes we used three sets of accessibilities (1993, 1998, and 2008) based on the output from the travel model used by MAG.

Model estimation and validation.
Given that UrbanSim is actually a group of models that communicate with each other through a data store, the estimation process involves separate calibration of each individual model. Most of the models are of a "discrete choice" nature and are estimated through nested logit regressions (e.g., household location choice model, developer choice model, and employment location choice model). The land price model is different from the previous set since it is the only model that is estimated through a linear regression procedure. The model parameters are derived with the help of external packages such as Limdep and SPSS.

Using the estimated parameters, two model configuration tables were generated for each model – the model specification table, and the model coefficient table. These tables are usually the last tables to be generated. Once these tables are populated with appropriate parameters, UrbanSim model runs can be accomplished. Figure 13 provides the 1990 and 2015 household location results for one UrbanSim run using the "business as usual" scenario.

Model validation is a crucial process for building confidence in the modeling results. For this paper, UrbanSim model is run from 1990 through 2000 and the simulated results are compared to the observed data to check the validity of the model. Practical constraints on creation of historical data for use in validation often preclude the feasibility of historical validation of this sort, but this remains one of the most informative ways to assess the model before putting it into

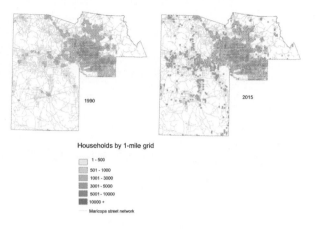

Households by 1-mile grid

☐ 1 - 500
☐ 501 - 1000
☐ 1001 - 3000
■ 3001 - 5000
■ 5001 - 10000
■ 10000 +

⋯ Maricopa street network

Fig. 13 Househode simulation using " business as usual" scenario

Table 3. Correlation of simulated to observed 2000 values

	Cell	TAZ
Employment	0.8	0.71
Households	0.76	0.66
Housing Units	0.79	0.64

operational use. The simulation results are compared to observed data at two units of geography. As seen in Table 3, the correlation between the simulated and observed is close to 80 percent at the level of the grid cell. However, this correlation is lower when a larger unit of geography such as the transportation analysis zone is considered.

Analysis of UrbanSim Scenarios with and without light rail.
The introduction of light rail in the Phoenix metropolitan area seems to increase the number of households in zones 1 and 2 when compared to a future without light rail. Between 2008 and 2015 the number of households in zones 1 and 2 increased 19 percent and 15 percent without light rail, respectively. Zone 3 also registers an increase in the number of households in this scenario by 6 percent. In contrast, the scenario with light rail assigns very slight changes to household numbers in zone 1, but significant increases in zone 2. The number of zone 2 households increases by 12 percentage points during that same period when compared to no light rail scenario. Figures 14 and 15 show the change in households by year for the two scenarios described above.

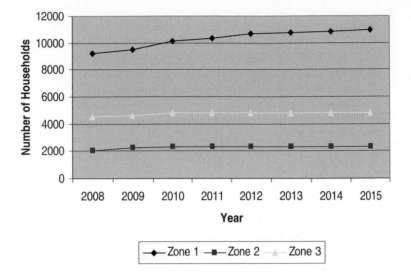

Fig. 14 Household simulation using "business as usual" scenario

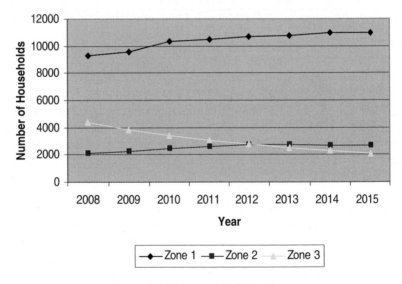

Fig. 15 Change in number of households by zone with light rail

A surprising result is noticed for zone 3, which includes large concentrations of high density student housing. The scenario with light rail seems to decrease the number of households in the seven years after the commencement of light rail in the Phoenix metropolitan area. Compared to the "no build" scenario, the introduction of light rail results in a decline of households by over 50 percent. Although the reduction in household density seems surprising, the model is behaving as expected given that capitalization of the amenity provided by light rail transit in home values may perhaps lead to new up-market developments that pushes out the lower-income student population and makes room for higher income families who prefer slightly larger quarters. This projected household transition becomes even more apparent when we examine the type of households who would prefer living adjacent to transit stops as predicted by UrbanSim.

Household transition due to light rail.
Characteristics of households in the three zones show different trends based on scenarios with and without the introduction of light rail transit in 2008. In this paper we report on two of the important characteristics of projected future households adjacent to light rail station areas ? income and race.

The three zones delineated for the study include, on average, low- to moderate-income households and the "no build" scenario does not change that overall character. Under the "no build" scenario, zone 1 registers the highest income of the three zones during the period of projection. Zone 2 remains the lowest in terms of average household incomes of the three zones. Both zones 1 and 2 show slight declines in real average incomes over the seven-year period of projection. Zone 3, however, registers significant decline in real average household income of about 8 percent during this period. This result changes dramatically in the scenario with light rail, especially for zone 3.

The scenario with light rail has significant yet differential impacts on zone 1 and zone 3. Households in zone 2, in contrast, are less likely to be of a different income group with or without light rail. Average household income in zone 1 is projected to decline significantly in the seven years after the initiation of light rail transit. In contrast, zone 3, which includes the student community around Arizona State University, is projected to scale up in average income levels during the same period. While households in zone 1 remain the highest in average incomes of the three zones without light rail, they give up that top position to households in zone 3 when light rail is introduced. Zone 2 households remain at the bottom in average income in either scenario.

With changes in household incomes, the racial ethnic composition of the households in the three zones also changes depending upon the introduction of light rail. In all but one scenario, the percentage of white households (as determined by the racial attribute of the head of household) decline from 2008 to 2015. White households comprise about 72 percent of all households in zones 1 and 2, and 64 percent in zone 3 in 2008. In the scenario without light rail, the highest decline in the percentage of white households is in zone 2 (6 percentage points) followed by zone 1 (3 percentage points) and zone 3 (1 percentage point). This decline is almost entirely at the expense of percentage growth of households in the "other" racial category. The "other" category is a residual category used in the U.S. Census for those individuals who do not choose among the dominant racial categories for various reasons including unwillingness to disclose or being of mixed races.

The racial make-up of the three zones seems to be very different in the scenario with light rail than the previous scenario. The decline of white households in zone 1 is now more pronounced (10 percentage points). However in zone 2, which had the largest decline of white households in the previous scenario, the percentage of white households now decline by only 4 percentage points. More surprisingly, percentage of white households in zone 3 actually trend up in the scenario with light rail by a significant 6 percentage points. In essence, zone 3 will be the most impacted area with the introduction of light rail partly due to gentrification.

3.2.2 The Final Analysis

The scenarios evaluated to test the impact of light rail on adjacent neighborhoods in the Phoenix metropolitan area show different impacts in different zones. The findings are mostly in line with the literature on transit and land use connections but also add some surprising caveats to this literature. While, as expected, zones 1 and 2 register slight increases in residential density over seven years since introducing light rail, household densities in zone 3 actually decline under this scenario. This result can be explained in light of current characteristics of zone 3 and its unique location. The household density in zone 3 is already among the highest in the state and includes a high percentage of student households. Given the income profile of this young student population, the housing available is mostly rental, aimed at low- to mid-market clients. In addition, this area is among the most "jobs rich" areas in the state being close to the fourth largest university in the U.S. and to downtown Tempe. Therefore, the perceived accessibility of this area is already high and the introduction of light rail transit provides the additional amenity that would make it more desirable to up-market clients.

The projected gentrification of zone 3 is especially unwelcome for the student population who would be gradually pushed out to areas farther from the university. Given this possible scenario, both the city of Tempe and Arizona State University will have to plan ahead for more affordable student housing in the future. The university has already embarked on an extended plan to increase on-campus student housing. The city also needs to closely monitor land use changes and real estate values in zone 3 and look for innovative approaches for developing as well as keeping affordable housing. Regardless, this area seems to be ripe for redevelopment and the introduction of light rail will perhaps jump start the process.

An important caveat to keep in mind is that simulation models are useful tools for understanding the interaction of contextual elements and decision-agents but they are limited in their capacity to anticipate processes that have no antecedents. This limitation is more pronounced in very long-range projections. The simulation results reported in this paper is well within the period in which projections can be justifiably made, given well verified models. However, the results should be treated as informational and not definitive since human social behavior changes through time due to adaptation and learning. Regardless, planning for the future requires us to anticipate it and the careful use of simulation and/or modeling tools is indispensable for this endeavor.

4 Next Steps

The challenge of the Digital Phoenix Project is to provide an integrated set of decision-making and visualization tools that will allow us to explore various options for developing the built environment in the Phoenix metropolitan area. In addition, these tools will address a range of research and policy questions about how the decisions we make now with regards to transportation options, land-use classifications, building designs, and urban landscapes among others will affect the future livability of this region. The project is unprecedented in scope and will undoubtedly offer several technical and theoretical challenges. The strategy so far has been to develop some strong expertise in a few areas in which there has been prior activity and some track record. By building upon existing research, we improve the viability of the project at least in terms of short-term deliverables. Regardless, the real contribution of the Digital Phoenix Project will be the integration of various disciplinary tools and theoretical frameworks in a manner such that policy and research questions that are transdisciplinary in scope can be addressed. This will be a significant contribution to scholarship as well as a critical and novel tool for decision making.

The team is focused on these integrative efforts, while, at the same time, developing expertise in specific tools and approaches. Given the long-term horizon, we have been ambitious and bold in our objectives. Regardless, we are aware of the importance of short term products to generate interest and visibility. We intend to create both.

We are embarking on several projects that will bring many of the project teams together to solve distinct problems. For example, the teams working on sustainability metrics and indicators of sprawl will engage with the team collating historical data about the evolution of Phoenix to provide sustainability indicators not just of the present and future scenarios but also of the past. Similarly, the team working on virtual reality engines of current downtown environment is being supported by travel demand models, which also provides critical information for urban simulation. Another project that has serious research challenges is developing 3-D visualizations of future scenarios that are typically mapped in 2-D space. We are undertaking research to develop algorithms for generating 3-D future virtual environments. We are also beginning to address questions at the neighborhood scale. One of the recent projects we have identified is the visualization of alternative design and density solutions for affordable housing. This project has immediate application to community decision making in a Decision Theater environment. Yet another project evaluates the consequences for energy use and heat island effects for different built environments. In essence, we have been able to attract new ideas and new contributions within our overarching vision for the Digital Phoenix project. Within the next year, we expect to make significant contributions to the knowledge of and tools for visualizing sustainable future environments.

References

Batty M (2005) Cities and complexity: understanding cities with cellular automata, agent-based models, and fractals. MIT Press, Cambridge, MA

Benenson I, Torrens P (2004) Geosimulation: automata-based modeling of urban phenomena, Wiley, Hoboken, NJ

Früh C, Zakhor A (2003) Constructing 3D city models by merging ground-based and airborne views. Proc. of Computer Vision and Pattern Recognition, Vol. 2, June

Guhathakurta S (2002) Urban modeling as storytelling: using simulation models as a narrative. Environment and Planning B 29:895–911

Guhathakurta S (2003) Integrated urban and environmental models: a survey of current applications and research, Springer, Heidelberg

Joshi H, Guhathakurta S, Konjevod G, Crittenden J, Li K (2006). Assessment of impacts of the light rail on urban growth in the Phoenix metropolitan region using UrbanSim modeling environment. Journal of Urban Technology 13 (2):91–111

Portugali J (2000) Self-organization and the city. Springer, Heidelberg

Waddell P (2002) UrbanSim: modeling urban development for land use, transportation and environmental planning. Journal of the American Planning Association 68 (3):297–314

IV
PhD-Theses

Estimating Residential Building Types from Demographic Information at a Neighborhood Scale

Ariane Middel

1 Introduction

There has been considerable interest of late in the generation of cityscapes in three dimensions, sometimes along time scales from some point in the past (Früh and Zakhor 2003; Shiode 2000; Parish and Müller 2001; Wonka et al. 2003). The interest has been driven by a number of developments including: 1) recently available computing technologies such as Google Earth and SketchUp; 2) the needs of wireless and cellular providers to track electromagnetic signals, and 3) the potential ability of urban designers and planners to communicate proposed buildings and spaces in a realistic mode to decision makers. Many cities such as London, Glasgow, Philadelphia, Los Angeles, and others have commissioned projects to develop virtual three-dimensional cityscapes. Other cities, such as Rome, have opted for reconstructing its form in virtual space at some point in the past. These virtual city models are becoming even more realistic with accurate depiction of street furniture, terrain, and street life. While the technology for modeling existing and past stages of cityscapes is becoming more sophisticated, the depiction of future neighborhoods and urban spaces, beyond what is deliberately designed, is still in its infancy. The ability to visualize how cities and urban areas may look 10 or 15 years from today, given specific tastes and technologies, communicates effectively the impacts of decisions made today on future environments.

Urban and regional planning agencies in most industrialized countries are now mandated to provide official projections of the level, intensity, and character of future urban growth. These projections are usually depicted spatially with the help of a Geographic Information System (GIS) platform. The modeling environment behind the GIS platform can be one of several available from commercial vendors (such as "What If?" or CommunityViz) or through university research labs (such as UrbanSim or CUFS). The output of these modeling environments is visualized in a

A. Middel
Decision Center for a Desert City, Global Institute of Sustainability, Arizona State University,
P.O. Box 878209, Tempe, AZ 85287-8209, USA, E-mail: ariane.middel@asu.edu

G. Steinebach et al. (eds.), *Visualizing Sustainable Planning*,
© Springer-Verlag Berlin Heidelberg 2009

two-dimensional space that is geographically aligned to a base map. A more real-istic built environment of the future visualized in three dimensions has rarely been attempted, mostly due to issues of empirical validity and computational complexity. In this paper, we develop an empirical model for generating three-dimensional future neighborhoods based on demographic characteristics. The intension is to provide an analytically appropriate technique to link demographic attributes with building and neighborhood typologies. In a subsequent paper, we show how these building typologies can be automatically generated by using database approaches and computational graphics.

2   Related Work

Among the various automated techniques to acquire spatial and geometric information about buildings, remote sensing is perhaps the most efficient. Tools and techniques are constantly being developed to automatically detect and reconstruct buildings from remotely sensed imagery for generating 3D city models (Brenner 2005) and to undertake building energy requirement calculations (Neidhart and Sester 2004).

Yet, the classical reconstruction approach does not work for extracting informa-tion on virtual or future urban environments. To overcome this problem, procedural modeling of buildings has been subject to active research lately. Procedural modeling is a computer graphics technique for algorithmically generating geometry from a predefined rule-set. Recent methods have been developed to procedurally generate architecture (Wonka et al. 2003), buildings (Müller et al. 2006), and large-scale cities (Parish and Müller 2001; Silveira and Musse 2006). These tech-niques generate realistic-looking urban environments, but to date they do not model realistic-growing environments in terms of predicted urban development. Until now, procedural models lack an underlying empirical framework for future urban land use, transportation, and environmental impacts as incorporated in the urban simulation software UrbanSim (Waddell 2002). UrbanSim is a sophisticated plan-ning tool for analyzing long-term effects of land-use and transportation policies. It provides a platform for generating different urban scenarios based on current trends and specified policy choices, e.g., to model low-density urban sprawl or to evaluate the sustainability of transportation plans (Joshi et al. 2006).

The challenge remains to map the 2D simulation results of land usage, density, and other significant socio-economical and statistical parameters into 3D space. For this purpose, building types and their spatial distribution have to be derived from the modeled output data. Standards for storing and representing thematic information on buildings in 3D city models are currently being developed (Kolbe et al. 2005), but none have so far been applied in a real context.

We propose a multinomial logit model (MNL) to estimate building types from predicted demographic data. MNL is widely used in social sciences and has a long tradition in the economics of consumer choice. Multinomial logit refers to the conditional discrete choice model, first introduced and most notably influenced by McFadden (McFadden 1973; McFadden 1976; McFadden 1978). Since then, a lot of research has been conducted in econometrics to develop multinomial logit models of residential location choice. John Quigley investigated consumers' qualitative choice behavior of residential location and building type in Pittsburgh, using a nested MNL (Quigley 1976), a generalization of McFadden's conditional logit model. Weisbrod assessed household location choice based on tradeoffs between accessibility and other housing and location characteristics (Weisbrod et al. 1980). Multinomial logit models were more recently applied to estimate the relationship between the overall level of housing prices and the mix of building types (Skaburskis 1999). Furthermore, discrete choice analysis was successfully adopted in transportation planning to examine travel demand (Ben-Akiva and Lerman 1985).

As noted above, multinomial logit analysis has a wide range of applications in behavioral assessment and categorical data analysis. In the remainder of this paper, we will present how MNL can be applied to estimate building type distributions at a neighborhood scale based on demographic household information.

3   Mapping the Real World

Let D denote a convex set of demographic household characteristics d, B denote a convex set of building types b and C be a set of contexts c. We assume that demographic household characteristics within a predefined context have an impact on the residents' choice of building types. Then, the mathematical mapping function Φ from the input sets D and C to the output set B can be derived by discrete choice modeling from the relationship $(D,C) \xrightarrow{\Phi} B$ with given input and output sets.

3.1   Context

The context vectors c are discretely time dependent and contain the social, legal, and political framework associated with the spatial location of the modeled

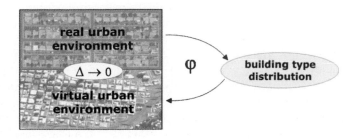

Fig. 1 Mapping

environment. The context can be used to calibrate the model and to generate different scenarios. In the following, C will be treated as exogenous and not part of the model. For the analysis, C is predetermined to match the current context of Maricopa County to simplify the model.

3.2   Demographics

In our model, the demographic vectors d store aggregated household information on demographic backgrounds from the US Population Census in 2000. Census data is accumulated statistical data on numbers of persons and selected social, economic, and financial characteristics. The demographic data at hand is at a Census block group level. A block group is comprised of Census blocks, the smallest geographic subdivision for which the Census Bureau tabulates data.

The following measures of block group characteristics are included into d for the subsequent analysis:

- population density
- median age
- average household size
- median household income
- average number of children per household
- average number of cars per household
- percentage of Hispanics and minorities
- distance to nearest highway
- average age of structures

3.3   Building Types

The building type vector b is a container for information on the physical characteristics of dwellings. In the current model, b is restricted to residential buildings, but future analyses will also incorporate commercial and industrial dwellings.

We derive the building type vectors from the Maricopa County Assessors data for 2000, which is an extensive database providing detailed property information at parcel level. We use ArcGIS to intersect the Assessors file with the Census data, to prepare and store the table as a Shapefile. According to the Primary Use Code, we define single family dwellings and apartments as the two main residential building type categories. The single family category is subclassified into lots with different sizes, the apartments are subdivided according to the number of housing units.

Altogether, we obtain nine different building type categories:

Apartments	Single family dwellings
• 2–24 units	• lot size <6,742 sq ft (XXS)
• 25–99 units	• 6,742 < lot size < 7,986 sq ft (X)
• >100 units	• 7,986 < lot size < 9,801 sq ft (S)
	• 9,801 < lot size < 12,705 sq ft (M)
	• 12,705 < lot size < 18,150 sq ft (L)
	• lot size > 18,150 sq ft (XL)

Figures 2–4 show aerial photographs of representative building types. After establishing the building type categories, we obtain a frequency count of every building type for every block group and transform the count data into percentages. The result is a building type composition where all building types add up to 100 percent in each block group.

4 Clustering

The building type data table was prepared on Census block group level. A block group is an aggregation of Census blocks and generally contains between 600 and 3,000 people. Due to the relatively coarse spatial resolution, most block groups

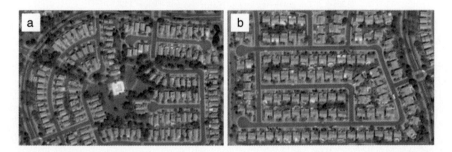

Fig. 2 Single family dwellings, very small lots (**a**), small lots (**b**)

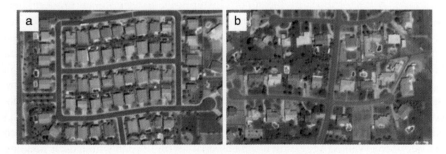

Fig. 3 Single family dwellings, medium lots (**a**), large lots (**b**)

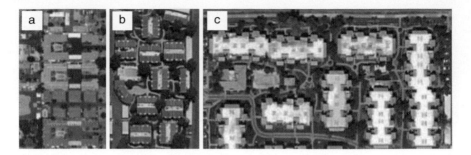

Fig. 4 Apartments, 5 units (**a**), 25–100 units (**b**), >100 units (**c**)

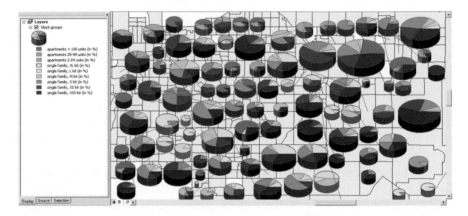

Fig. 5 Percentages of different building types in each block group

assemble a mix of different building types rather than being a homogeneous composition of dwellings (see Fig. 5). Consequently, we have to find similar building type distributions and classify them into different typical neighborhood categories. To build these categories, the percentages of building types in each block group are calculated from frequency count data and then grouped together into classes of similar neighborhoods by means of clustering. Clustering is an unsupervised learning technique used in statistical data analysis to determine the inherent grouping in a collection of unclassified data. For an overview of classical data mining techniques and a detailed description of different clustering methods see (Tan et al. 2005).

We use a standard k-means algorithm to find significant neighborhood patterns in the building type data table. K-means iteratively partitions m data points $x_1, x_2, \ldots x_m$ into a user-specified number of k clusters $c_1, c_2, \ldots c_k$, assigning all data points to their closest cluster centroid. In our case, the data to be clustered is a set of real-valued vectors in n-dimensional feature-space where each building type adds a new dimension. Distances between those vectors are defined by the Euclidean metric.

Table 1 Final cluster centers

	Cluster					
	1	2	3	4	5	6
apartments 2–24 units (in %)	3.6	39.2	2.0	2.2	3.4	2.8
apartments 25–99 units (in %)	5.2	9.5	0.1	0.9	1.0	1.2
apartments >100 units (in %)	69.9	4.4	0.0	3.4	4.0	2.7
single family, XXS lots (in %)	5.8	11.5	2.8	8.4	52.2	5.5
single family, XS lots (in %)	5.9	15.7	2.1	30.6	23.3	5.3
single family, S lots (in %)	4.1	8.6	2.1	38.6	10.0	16.1
single family, M lots (in %)	1.8	4.2	2.8	10.8	4.3	36.4
single family, L lots (in %)	1.0	2.5	6.4	4.3	1.8	19.3
single family, XL lots (in %)	1.1	2.6	83.0	1.9	1.6	12.1

The *a priori* choice of the optimal number of clusters k is essential for reasonable clustering results. Forming homogeneous clusters is especially difficult for high-cardinality data. Choosing a large k will reduce the impact of noise on the categorization, but will also result in fuzzier cluster boundaries. The challenge is to create homogeneous clusters and at the same time minimize k.

Table 1 shows the clustering result of the k-means algorithm after 14 iterations, presenting the centroids for every cluster. The optimal number of clusters $k = 6$ was determined analytically. From the most representative cluster prototypes we see that apartments with over 100 units and single family dwellings with very large lots clearly stick out as the prominent building types in clusters no. 1 and 3. The typical building type distributions of the other clusters are more balanced with variable dwelling types.

After finding the cluster centroids, each block group in the data set is assigned to the nearest cluster centroid considering the dwelling type distribution and thus obtains a neighborhood category for the following prediction.

5   Multinomial Logit – MNL

To estimate the distribution of building types at a block group scale, we use regression analysis. Regression maps the values from a predictor variable in such a way that the prediction error is minimized. The analysis to be performed involves more than one predictor, since the input data, discussed in section 3.2, is given as 2040 multi-dimensional vectors having one dimension for every demographic variable. In order to investigate the effect of more than one independent variable on the discrete outcome, we use multinomial logistic regression. This kind of regression can handle several predictors as input variables as well as polytomous response variables with more than two output categories. The process of predicting values of multilevel responses with unordered qualitative categorical outcomes is described next (Powers and Xie 1999).

Consider the polytomous outcomes y_i for a dependent variable with J categories where i denotes the ith respondent. Let the response probability P_{ij} represent the probability of a particular outcome, thus the chances that the $i$ths respondent falls into category j. Consider that x_i is the vector of predictors storing values for the independent variables and $\boldsymbol{\beta}$ the regression parameter vector. Then, the linear relationship

$$z_{ij} = x'_i \boldsymbol{\beta}_j = \sum_{k=0}^{K} \beta_{jk} x_{ik} = \alpha_j + \sum_{k=1}^{K} \beta_{jk} x_{ik}$$

renders the parameters of the model estimable, and we can calculate the set of coefficients β_{jk} which correspond to the covariates x_{ik} of the response probabilities. For the outcome y_i, the probability of selecting j is:

$$P(y_i = j | x_i) = P_{ij} = \frac{e^{z_{ij}}}{\sum_{j=1}^{J} e^{z_{ij}}} = \frac{e^{x'_i \beta_j}}{\sum_{j=1}^{J} e^{x'_i \beta_j}} = \frac{e^{x'_i \beta_j}}{1 + \sum_{j=2}^{J} e^{x'_i \beta_j}}$$

This probability statement has the constraint that all probabilities have to sum up to 1:

$$\sum_{j=1}^{J} P_{ij} = 1$$

The multinomial logistic regression is a generalized discrete choice model for nominal response variables (Agresti 2002). Discrete-choice multinomial logit models are widely used in economics and differ from the standard model in a way that the explanatory variables vary not only by outcome but also by individual. It is assumed that the individual has preferences over a set of alternatives, e.g. travel modes, and chooses the alternative which maximizes utility:

$$y_i = j \text{ if } u_{ij} \geq \max (u_{i1}, ..., u_{ij}) \qquad u_{ij} = r'_{ij} \beta + \varepsilon_{ij}$$

$r'_{ij} \beta$ is the systematic component of alternative j that considers the characteristics of the choice as well as the preferences of the individual and ε_{ij} is a stochastic term.

The multinomial logistic model simultaneously describes log odds for all $\binom{J}{2}$ category pairs. The log-odds of membership in one category of the dependents versus an arbitrary baseline category, normally the first category, are fitted as a linear function of covariates x_i:

$$\log \left(\frac{P_{ij}}{P_{i1}} \right) = x'_i \beta_j$$

The odds between two arbitrary categories j and j' are calculated as follows:

$$\frac{P_{ij}}{P_{ij'}} = e^{x_i\,(\beta_j - \beta_{j'})}$$

The regression parameters are estimated using maximum likelihood. This is a stable calculation, since the log-likelihood of the probabilities here is convex.

6 Empirical Analysis and Results

In section 4, we derived neighborhood categories with typical building type distributions. These categories are used as dependents x_i in our multinomial regression model. The six outcomes of this building distribution variable are interpreted according to the cluster centers (see table 1)

1. mainly apartments
2. balance between apartments and single family, majority of apartments <25 units
3. mainly single family, XL lots
4. mainly single family, S and XS lots
5. mainly single family, XS and XXS lots
6. mainly single family, majority of M lots

Table 2 shows the number and percentage of block groups in every cluster that are included into the analysis.

The demographic vectors d of aggregated household information on block group level, introduced in section 3.2, are the explanatory variables considered in the estimation process. To predict the outcome categories in the multinomial logit model, we incorporate 2040 block groups of Maricopa County with a total number of 1,115,570 households. The rounded average household size per block group is 3 persons, and the average median income per household is $48,745. About 17% of the block groups have a majority of Hispanics, whereas 75% of the block groups have mainly white population. The number of children per household in every block group averages 0.7 with a rounded number of one car per household.

Table 2 Case processing summary

		N	Marginal percentage
	1	283	13.9%
	2	327	16.0%
	3	117	5.7%
	4	539	26.4%
Cluster number of case	5	555	27.2%
	6	219	10.7%
Total		2040	

After building the MNL model with the described input data, we validate our model to evaluate whether it approximates the behavior of our data in a reasonable way. Below, we will assess the fit of our model to the data with different statistical tests.

6.1 Goodness of Fit Test

Pearson and Deviance are the most prevalent goodness-of-fit statistics used to validate the model in multinomial logistic regression. They test the null hypothesis that the model adequately fits the data. In our logit model, we have many predictors with continuous values as covariates causing many subpopulations with zero frequencies. Because of the many cells with expected zero values, the test statistics lack large sample properties and a dependable goodness-of-fit test is not provided.

6.2 Pseudo R^2

In multinomial logistic regression, a direct analog to the R-Squared statistic as used in ordinary linear regression does not exist. R^2 cannot be computed, since R^2 measures the variability in the dependents, but the variance of a categorical predictor is a function of the variable's frequency distribution. To summarize the strength of the association between the dependent and independent variables, MNL uses pseudo R^2 statistics which are designed to have similar characteristics to the R^2 statistic. Regression output Table 3 shows three pseudo R^2 estimates: Cox and Snell R^2, Nagelkerke's R^2, and McFadden's R^2. Larger values between 0 and 1 indicate a better explanation of the variation by the model, which means that our model performs reasonably well.

6.3 Likelihood-Ratio Tests

Before examining individual coefficients, we will assess the significance of the MNL by showing that the model fits the data better than a null model. The overall test of the null hypotheses that the regression coefficients β for all of the variables x in the model are 0 is called likelihood ratio test. Likelihood denotes the probability that the estimated values of the dependent may be predicted from the independents.

Table 3 Pseudo R-Square

Cox and Snell	0.736
Nagelkerke	0.763
McFadden	0.398

Table 4 Model fitting information

Model	−2 Log Likelihood	Chi-Square	df	Sig.
Intercept only	6841.390			
Final	4121.008	2720.382	55	0.000

Table 5 Likelihood ratio statistics, reduced model

Effect	Model Fitting Criteria	Likelihood Ratio Tests		
	−2 Log Likeli-hood of Reduced Model	Chi-Square	df	Sig.
intercept	4464.185	343.178	5	0.000
population density	4368.175	247.168	5	0.000
median age	4216.168	95.160	5	0.000
median household income	4257.356	136.348	5	0.000
average #children per house-hold	4128.715	7.707	5	0.173
average #cars per household	4128.971	7.963	5	0.158
minorities in %	4150.439	29.432	5	0.000
Hispanics in %	4139.113	18.106	5	0.003
city	4132.266	11.259	5	0.046
average household size	4242.518	121.510	5	0.000
distance to nearest highway	4156.048	35.041	5	0.000
average age of structure built	4432.104	311.096	5	0.000

The likelihood ratio is a function of log likelihood and makes a statement about the significance of the unexplained variance in the outcome. As shown in Table 4, the chi-square distributed difference between the −2 log-likelihood values for the null hypothesis and the final model has an observed significance level of 0.000. Therefore, we can reject the null hypothesis that the MNL model without predictors performs as well as the model with predictors.

The results prove strong support of the overall model significance, but the previous test does not assure that every predictor variable is significant for the prediction.

In order to test individual model parameters, we have to check the contribution of each predictor to the model separately. Therefore, we create a reduced model by omitting one independent variable at a time and use a likelihood ratio test to analyze the differences in −2 likelihoods between the overall model and the nested model. This equals a test whether the coefficient for the omitted effect can be treated as zero since the effect does not have an influence on the regression. Table 5 shows that the test clearly rejects the null hypothesis for almost all independents included in the analysis. Yet, the average number of cars and children per household do not contribute to the model at a very high level of significance (level of significance >0.05). For this reason, the variables will be excluded from the MNL model.

6.4   Classification

An indicator for how well the multinomial logistic regression model predicts categories of the polytomous dependent variable is the so-called classification table. Table 6 shows the classification table for our MNL model. This *JxJ* table crosstabulates observed categories with predicted categories and helps measuring correct and incorrect estimates. The diagonals contain correct predictions, whereas cells off the diagonal are falsely predicted. For example, 198 of the 283 block groups observed to be in neighborhood category 1 (mainly apartments) were classified correctly.

The table shows good results in terms of correct predictions. In most of the cases, the percentage of correctly predicted categories exceeds 60 percent. The null model, which classifies all cases according to the modal category, classifies correctly for only 13.9% of the cases (compare table 2). Furthermore, the classification table proves that our MNL has homoscedasticity, since the percentage of correct predictions is approximately the same for every row.

6.5   Parameter Estimates

The model parameters of our multinomial logistic regression are summarized in Table 7. For every cluster except the reference category, we get estimated logit coefficients β associated with every predictor as well as a value for the intercept. The *(J-1)* logits β can be used in prediction equations to generate logistic scores and thus are the key to predicting future building types.

The algebraic sign of the coefficients determines the effect of each predictor on the model. Positive parameter estimates like the average household size increase the likelihood of the response category with respect to reference category 1. Responses with significant negative coefficients as the percentage of minorities reduce the likelihood of that category. In general, the effect of the predictors is strongest for category 3 versus 1 and weakest for category 2 versus the reference category. The

Table 6 Classification table

Observed		Predicted						
		1	2	3	4	5	6	Percent correct
	1	198	43	1	12	27	2	69.96%
	2	39	216	4	23	43	2	66.06%
	3	1	3	86	6	8	13	73.50%
	4	15	32	7	331	129	25	61.41%
	5	27	58	8	151	301	10	54.23%
	6	2	7	16	84	17	93	42.47%
Overall percentage		13.82%	17.60%	5.98%	29.75%	25.74%	7.11%	60.05%

Table 7 Parameter estimates

Cluster number of case		β	Std. Error	df	Sig.	Exp(β)
2	Intercept	−8.9665	1.3334	1	0.0000	
	pop_norm	−0.2266	0.0646	1	0.0005	0.7973
	medianag	0.0504	0.0153	1	0.0010	1.0517
	medhhinc	0.0000	0.0000	1	0.0015	1.0000
	minor_ph	-0.0639	0.0164	1	0.0001	0.9381
	hispa_ph	0.0270	0.0177	1	0.1281	1.0274
	avehhsiz	3.2540	0.3504	1	0.0000	25.8931
	distance	0.0001	0.0001	1	0.2865	1.0001
	age	0.0971	0.0092	1	0.0000	1.1020
3	Intercept	−32.8111	2.6217	1	0.0000	
	pop_norm	−4.0936	0.3705	1	0.0000	0.0167
	medianag	0.2142	0.0271	1	0.0000	1.2389
	medhhinc	0.0001	0.0000	1	0.0000	1.0001
	minor_ph	−0.2323	0.0532	1	0.0000	0.7927
	hispa_ph	0.1137	0.0539	1	0.0348	1.1204
	avehhsiz	8.5353	0.6094	1	0.0000	5091.1159
	distance	0.0002	0.0001	1	0.0004	1.0002
	age	0.1713	0.0167	1	0.0000	1.1868
4	Intercept	−21.1080	1.4310	1	0.0000	
	pop_norm	−0.3909	0.0864	1	0.0000	0.6764
	medianag	0.1576	0.0151	1	0.0000	1.1706
	medhhinc	0.0001	0.0000	1	0.0000	1.0001
	minor_ph	−0.0564	0.0179	1	0.0016	0.9451
	hispa_ph	−0.0305	0.0205	1	0.1357	0.9699
	avehhsiz	5.9174	0.3884	1	0.0000	371.4462
	distance	0.0000	0.0001	1	0.9594	1.0000
	age	0.0743	0.0092	1	0.0000	1.0771
5	Intercept	−16.6966	1.3311	1	0.0000	
	pop_norm	−0.3195	0.0725	1	0.0000	0.7265
	medianag	0.1195	0.0150	1	0.0000	1.1269
	medhhinc	0.0001	0.0000	1	0.0000	1.0001
	minor_ph	−0.0387	0.0157	1	0.0137	0.9620
	hispa_ph	0.0004	0.0179	1	0.9809	1.0004
	avehhsiz	4.8590	0.3622	1	0.0000	128.8943
	distance	0.0001	0.0001	1	0.3375	1.0001
	age	0.0251	0.0087	1	0.0037	1.0254
6	Intercept	−25.7086	1.8542	1	0.0000	
	pop_norm	−1.1585	0.1451	1	0.0000	0.3139
	medianag	0.1579	0.0190	1	0.0000	1.1710
	medhhinc	0.0001	0.0000	1	0.0000	1.0001
	minor_ph	−0.0750	0.0301	1	0.0127	0.9277
	hispa_ph	−0.0373	0.0349	1	0.2852	0.9634
	avehhsiz	6.2972	0.4645	1	0.0000	543.0533
	distance	0.0002	0.0001	1	0.0082	1.0002
	age	0.1334	0.0118	1	0.0000	1.1427

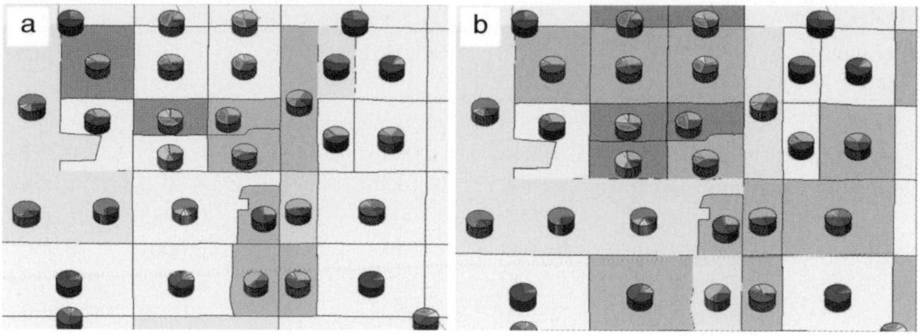

Fig. 6 Predicted vs. observed neighborhood categories

Fig. 7 Difference picture of correct vs. incorrect predictions

table also shows the standard error of the coefficients and the odds ratio, labeled as Exp(β). Looking at the significance of the explanatory variables, average household size, median age, average age of structure built, and population density are highly significant for the prediction. The significance level of the variable depicting the percentage of minorities varies but shows an overall significance whereas the percentage of Hispanics per block group does not seem to be important for the prediction at all. Finally, the distance to the nearest highway is not significant for three of the categories, but is very important for predicting categories 3 and 6.

Figure 6 shows a comparative visualization of the predicted and observed categories for sample block groups in Maricopa County. The prediction is based on the logit coefficients calculated earlier. Additionally, a difference picture of correctly and incorrectly predicted categories can be seen in Fig. 7. Here, block groups with matching response and observed category are colored blue, whereas neighborhoods with mismatching categories appear in red. Both visualizations affirm the prediction rate of about 60 percent as stated in the classification table (see section 6.4).

7   Conclusions and Future Work

In this paper, we presented a statistical method for estimating residential building types on a neighborhood scale from demographic data. The mapping between dwelling types and household characteristics was realized with multinomial logistic regression. Clusters with typical building type distributions were formed by k-means to establish nominal categories for the regression model. Subsequently, the log odds of the clustered neighborhood category predictors were modeled as a linear function of the categories' covariates in the estimation process. The results of the MNL model were tested for data fit significance. All tests provided strong support that the model fits the data reasonably well and that logit regression is a coherent framework for assessing the relationship between demographic characteristics and the neighborhoods people live in.

An important basic assumption within the MNL framework is the IIA assumption, the independence of irrelevant alternatives. Multinomial logistic regression assumes that the odds of predicting one outcome over any other outcome are not dependent on the number or characteristics of the other responses. If the IIA assumption does not hold, the logistic regression will yield biased estimates because of correlated error terms. In our model, we predict different neighborhood categories rather than distinct dwelling types due to the coarse block group resolution. Our clustering produces typical building type distribution categories that are as distinct as possible. Consequently, the predictors are as uncorrelated as possible.

To fully satisfy independence across the model's response variables, we will implement a nested logit model. For the nested MNL, we will use the synthesized household table on demographics from UrbanSim to guarantee homogeneous building types on a grid cell level (150×150 m). Further research will also incorporate an extension of the model to estimate industrial and commercial building types.

After refining the model we will begin visualizing the predicted building types resulting from future demographic data derived from UrbanSim output. This allows for a better visualization of different planning scenarios in three dimensions, which will improve the perception of density and sky-view factors at the neighborhood scale. Several techniques are being explored to generate the three-dimensional building objects including procedural or template-based database methods. This next stage of our research, together with the empirical models described in this document, will provide a unique tool for visualizing future environments in realistic urban contexts.

This work was supported by the German Science Foundation (DFG, grant number 1131) as part of the International Graduate School (IRTG) in Kaiserslautern on "Visualization of Large and Unstructured Data Sets. Applications in Geospatial Planning, Modelling, and Engineering."

Bibliography

Agresti A (2002) Categorical data analysis. Wiley, New York

Ben-Akiva M E, Lerman S R (1985) Discrete choice analysis: theory and application to travel demand. MIT Press, Cambridge, MA

Brenner C (2005) Building reconstruction from images and laser scanning. International Journal of Applied Earth Observation and Geoinformation 6(3–4):187–198

Früh C, Zakhor A (2003) Constructing 3D city models by merging ground-based and airborne views. Proc. of computer vision and pattern recognition, Vol. 2

Joshi H, Guhathakurta S, Konjevod G, Crittenden J, and Li K (2006) Simulating impact of light rail on urban growth in Phoenix: an application of urbansim modeling environment. Proceedings of the 2006 international conference on digital government research, ACM Press, San Diego, pp 135–141

Joshi H, Guhathakurta S, Konjevod G, Crittenden J, and Li K (2006) Assessment of impacts of the light rail on urban growth in the Phoenix metropolitan region using UrbanSim modeling environment. J of Urban Technology 13(2):91–111

Kolbe, T H, Gröger G, and Plümer L (2005) CityGML – Interoperable access to 3D city models. Proceedings of the int. symposium on geo-information for disaster management, Delft

McFadden D (1973) Conditional logit analysis of qualitative choice behavior. In: Zarembka, P. Frontiers in econometrics. Academic Press, New York, Berkeley

McFadden D (1976) Properties of the multinomial logit (MNL) model. Urban Travel Demand Forecasting Project, Institute of Transportation Studies, Berkeley

McFadden D (1978) Modelling the choice of residential location. Institute of Transportation Studies University of California, Berkeley

Müller P, Wonka P, Haegler S, Ulmer A, and Gool L V (2006) Procedural modeling of buildings. ACM Trans Graph 25(3): 614–623

Neidhart H, Sester M (2004) Identifying building types and building clusters using 3-D laser scanning and GIS-data. In: Geo-Imagery Bridging Continents, XXth ISPRS Congress, Istanbul, pp 715–720

Parish Y I H, Müller P (2001) Procedural modeling of cities. Proceedings of the 28th Annual Conference on Computer Graphics and Interactive Techniques, ACM Press, pp 301–308

Powers D A, Xie Y (1999) Statistical methods for categorical data analysis. Academic Press, New York, Berkeley

Quigley J M (1976) Housing demand in the short run: an analysis of polytomous choice. Explorations in Economic Research 3(1): 3

Shiode N (2000) Urban planning, information technology, and cyberspace. J of Urban Technology 7 (2):105–126

Silveira L G D, Musse S R (2006) Real-time generation of populated virtual cities. Proceedings of the ACM symposium on virtual reality software and technology. ACM Press, Limassol, Cyprus pp 155–164

Skaburskis A (1999) Modelling the choice of tenure and building type. Urban Studies 36(13): 2199–2215

Tan PN, Steinbach M, Kumar V (2005) Introduction to data mining. Pearson Addison Wesley, Boston

Waddell P (2002) UrbanSim: modeling urban development for land use, transportation and environmental planning. J of the Am Planning Assoc 68 (3):297–314

Weisbrod G, Ben-Akiva M, and Lerman S (1980) Tradeoffs in residential location decisions: transportation versus other factors. Transportation Policy and Decision-Making 1 (1):13–16.

Wonka P, Wimmer M, Sillion F, and Ribarsky W (2003) Instant architecture. ACM SIGGRAPH 2003 Papers. ACM Press, San Diego

Visualization of Sustainability Indicators: A Conceptual Framework

Ray Quay and Khanin Hutanuwatr

Introduction

Indicators are an effective tool in providing an assessment of sustainability at various spatial, functional, and temporal scales. However, there has been little research on indicator organization or visualization techniques that could be used to convey sustainability information in a way that meets the needs of public decision makers. This article proposes a theoretical organization and visualization framework that provides a high degree of flexibility for conveying sustainability indicator data that decision makers may need to assess a wide range of urban and regional issues.

Sustainability is a complicated concept spanning such a broad range of disciplines that few individuals can be experts in all aspects of sustainability. The common response is to focus on a particular scale or subject. However, if sustainability is to be successfully inserted into public decision making it will need to be one of many factors (EUROCITIES 2004, 6) relevant to a wide range of issues on which decision makers are focused (Clark 2003; Nyerges 2002). Given the complexity of sustainability information decisions makers will need it simplified to a manageable level useful for consideration along with other factors of importance (Forester 1989; Klein 1998; Lindblom 1995).

Indicators are one tool that can be used to help simplify sustainability issues (Gudmundsson 2003). In general most sustainability indicator models report information across three basic scales: topical, spatial, and temporal. The interests and needs of decision makers can vary across these scales based on the nature of the various issues they are evaluating (Prescott-Allen 2001). City council members may be interested in indicators at various spatial scales, such as the city as whole, individual neighborhoods, or perhaps a single building. They may also only be interested in specific topical areas, such as heat island effect, water and energy efficiency, air quality, and resource utilization. The scale of their interest may even

R. Quay(✉)
College of Design, Arizona State University, PO Box 872105, Tempe, AZ 85287-2105, USA,
E-mail: ray.quay@asu.edu, khutanuw@asu.edu

G. Steinebach et al. (eds.), *Visualizing Sustainable Planning*,
© Springer-Verlag Berlin Heidelberg 2009

change over the course of the issue discussion. Effectively delivering sustainability information to decision makers will require visualization techniques that allow for easy exploration of simple sustainability indicators at these various topical, spatial, and temporal scales (Clark 2003; Gallopín 2004; Nyerges 2001). Thus, there is a great need for flexibility in organizing and visualizing sustainability indicators.

Most models for sustainability indicators use two basic concepts for organization: 1) measures used directly or in some normalized fashion to create indicators, and 2) indicators organized into categories, usually hierarchical, based on some topical or systems classification scheme. In some cases, these indices are combined to create a higher-level topical index, and so on. However, there is little consistency among these models in how indicators are developed or aggregated functionally, spatially, or temporally. Each model creates its own standard classification, weighting, and aggregation standards.

Though there is a rich set of systems for defining sustainability indicators, there has been less work on general visualization of these indicators (Williams 2004). Each model is typically presented with one method of reporting such as tables, graphs, or maps. A few general visualization tools have been proposed, such as the Dashboard of Sustainability (Consultative Group on Sustainable Development Indicators 2006) and the Epsilon Project (Jolliet et al. 2003). However these proposals are either focused on a specific data set, do not address data organization to facilitate visualization, or do not provide a simple and easily understood interface to allow decision makers or policy analysts to browse the data in a manner that meets their particularly spatial, topical, or temporal scope.

A Proposed Theoretical Framework for Visualization of Sustainability Indicators (SI)

Though there are numerous models for defining and organizing sustainability indicators (SI) (SRP 2004; Central Texas Sustainability Indicators Project 2004; Venetoulis et al. 2004; Wackernagel et al. 1997), some published work defining methods of weighting and aggregation (Nardo et al. 2005), and some published work on developing visualization methods and tools (Consultative Group on Sustainable Development Indicators 2006; Williams 2004) there has been little published on a theoretical framework for organization and visualization of SI. This is a critical need. Sustainability is of increasing importance, and the public as well as the business and political communities are eager to incorporate sustainable practices into their daily lives and professional practices. With today's vast information resources, a lack of information will not be a barrier to this effort. Rather, the challenge will lie in trying to manage an overwhelming amount of information. Thus sustainability systems that present large amounts of data in an easily accessed and understood format are needed.

This paper proposes a theoretical framework of data organization and visualization for sustainability indicators to meet these anticipated information needs.

This framework has three components: 1) an organizational system that defines indicator data based on hierarchies of topical, spatial, and temporal attributes and methods for aggregation; 2) visual techniques to display indicator data in various views that either are topically, spatially, or temporally focused or defined; and 3) a simple and intuitive visual framework that would allow the decision maker to easily explore sustainability indicators and analysis through various views and scales.

Organization of Indicators in a Multi-Dimensional Hierarchical Data Structure and Methods of Aggregation

The calculation of an indicator is typically based on some rule that defines what data is used, what calculation is used (if any) to create a metric, and what normalization is done (if any) to create a numeric index or indicator. In addition to this numeric, these indicators have attributes of topic, space, and time. For example, an indicator value created from CO_2 levels is an indicator that describes some aspect of air quality. It is collected from a specific place and time thus can be described by topic, space, and time.

These attributes of scope (spatial, topical, and temporal) are not static, but can be described as part of a hierarchical structure. Space can be defined as a single parcel of land or aggregated into a single block, a square mile, a city, a state, a nation, a hemisphere or the planet as a whole. Topic can be described as a single measurement, such as CO_2 levels, combined with others to describe green house gases, combined with other measures (Particulates etc) to describe air quality, or as part of a summary of environmental health or general global well being. Time can be described as the present moment or a particular day, month, year, decade, or century or as the difference between two points in time.

Though these hierarchy schemes will be used to organize data, they can be described independent of the indicator data itself with defined methods of aggregation and summary of data based on the structure of the hierarchy. Each level of the hierarchy represents a summary of all the values in the level below it. This data summarization of indicators from one level to another is based on rules of how the indicators or data from lower levels are aggregated at the higher level. These rules will be different for each of the scope-based hierarchical schemes. For example, the aggregation of the various air quality indicators to a single air quality index may be done by simply averaging the values. However, in a spatial hierarchy, an air quality index of several countries may be aggregated to a region by weighting each countries' air quality index by the size of the country. Or in a temporal hierarchy a human welfare indicator for a particular decade may be weighted by the population of each country in each of the years of the decade. Thus within the hierarchical indicator scheme, rules must be developed for how each indicator is aggregated.

Visualization of Sustainability Indicators

Under this concept of hierarchical SI organization, the visualization of the SI database would be based on specific spatial, topical, and temporal levels of interest. Visualization of hierarchical data is a well explored field of visual analytics (Baker and Bushell 1995; Bots et al. 1999; Brandes et al. 2003; Chen 2004; Crampton 2001; DiBattista et al. 1994; Hallisey 2005; Jern and Franzén 2006; Muller and Schumann 2003; Robertson et al. 1991; Shneiderman 1992; Thomas and Cook 2005; Tufte 1990, 1997) and can easily be applied to the visualization of SI data.

These techniques represent two important functions: 1) an interactive method to allow navigation through hierarchies to find specific information based on a topical, spatial, and/or temporal scope, and 2) methods for visual representations of hierarchical data to allow the brain's pattern recognition abilities to analyze the data. For example, various types of network graphs have been used to create interactive systems that can be used to navigate hierarchies. Probably the most well known of these is the Windows File Explorer. But other methods are emerging as well such as Tree Maps (Shneiderman 1992) and Time Rivers (Muller and Schumann 2003). Tree Maps are well suited for visualizing and browsing very large data sets. Figure 1

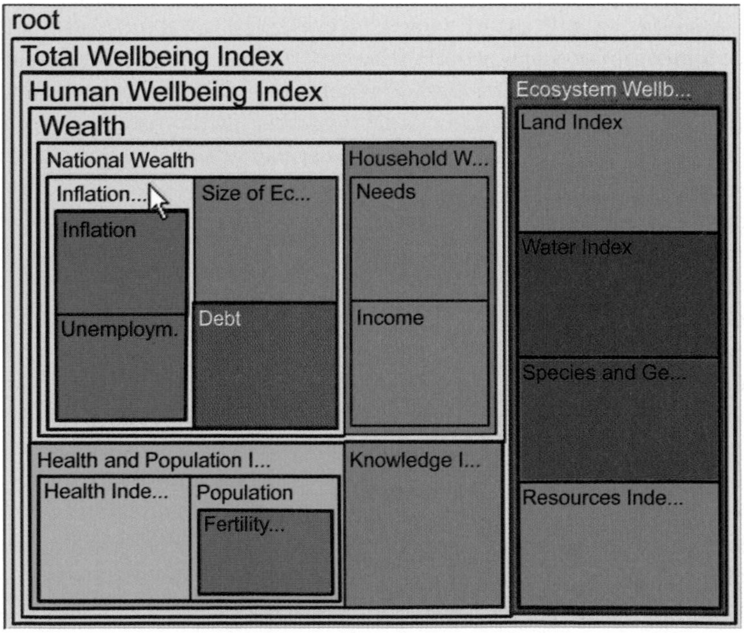

Fig. 1 Tree map of SI from well being of nations

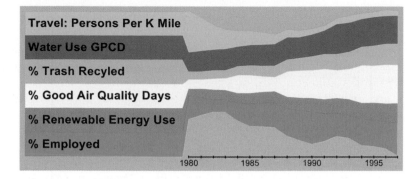

Fig. 2 Time river of SI from sustainable Seattle 1998

Float shows an example of SI data displayed in a Tree Map. Time Rivers are a method for displaying time series data. Figure 2 shows a Time River for SI data from Seattle's 1998 report.

A Framework for a Sustainability Browser

A key feature of this proposal is the organization of SI data in a framework of multiple hierarchies – spatial, topical, and temporal. What will be needed is a browser that will allow a user to navigate these multiple hierarchies to find the SI data of interest and select the visualization techniques that provides the best view of the data for a specific decision or issue.

Such a SI browser would have a user interface that allows the selection of data based on each scope hierarchy. Such an interface would provide hierarchal navigation frames that the user would use to browse the data based on single or multiple attributes selected from the hierarchies, and main display frame where selected SI data would be displayed using some visualization method. Figure 3 provides an overview of this kind of interface. Conceptual frameworks and prototypes are being developed on this basis and are located at www.public.asu.edu/~mcquay/sivp

Figure 4 Float shows this basic concept for a topical view using indicators from the Wellbeing of Nations. In this example Total Wellbeing with a breakdown of the two basic indicators that are used to calculate the total well being indicator, Human Wellbeing and Ecosystems Wellbeing, are displayed in an expanding list. The user could see more specific topical information by expanding a level, in this case Human Wellbeing. As each level is expanded, all of the indicators for this level and their values would be displayed. Each view would be based on a particular temporal and spatial reference, also selected using an interactive visualization of these hierarchies.

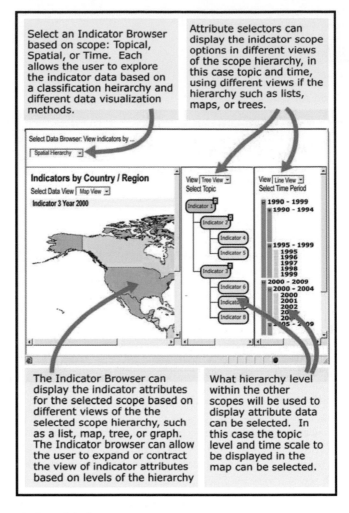

Fig. 3 Conceptual user interface

Tables 1 through 3 provide a description of topical, spatial, and temporal views in terms of organization and visualization.

Future Research

This article provides a logical visualization framework for decision makers to explore sustainability indicators and the measures behind them. However most of the current sustainability indicator systems do not have sufficient organization and/ or data to utilize the full capabilities of such a framework. Six key areas of research are needed.

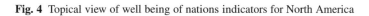

Indicator Topic	Value	Low (1-100) High
−Total Wellbeing Index	48.5	‖‖‖▨‖‖‖
+Human Wellbeing Index	65.3	‖‖‖‖■‖‖
+Ecosystem Wellbeing	31.7	‖■‖‖‖‖‖

Indicator Topic	Value	Low (1-100) High
−Total Wellbeing Index	48.5	‖‖‖▨‖‖‖
−Human Wellbeing Index	65.3	‖‖‖‖■‖‖
▯ Health and Population Index	75.3	‖‖‖‖‖■‖
▯ Wealth Index	69.3	‖‖‖‖■‖‖
+ Community Index(Freedom and Peace)	63.5	‖‖‖▨‖‖‖
▯ Knowledge Index	51.0	‖‖‖▨‖‖‖
▯ Equity Index	49.3	‖‖‖▨‖‖‖
▯ Human Development Index	88.0	‖‖‖‖‖‖■
+Ecosystem Wellbeing	31.7	‖■‖‖‖‖‖

Fig. 4 Topical view of well being of nations indicators for North America

Table 1 Topical views of sustainability indicator

Concepts	Structure	Detail
Organization	Hierarchical Category Trees	Hierarchically structured topics with aggregation rules.
	Scales	Data for each topic unit would be reported for each spatial and temporal scale or be based on the aggregation rules for one or all scales (spatial, topical, temporal).
Schemes	System	Example: Prescott-Allen's *The Wellbeing of Nations*
	Principles	Example: Natural Step http://www.naturalstep.org/, United Nations Rio Declaration, American Planning Association's Policy Guide Planning for Sustainability, World Summit on Sustainable Development
Display	Data Display	Data could be presented as a list, tree map, or graph view. In these views SI values for all topical units at the selected topical hierarchy level would be based on the selected Spatial and Temporal scales. List views would list topics and SI values in numeric and/or iconic formats. Tree Map views would display SI values in a tree map with each indicator map size based on their % contribution to the indicator of the higher level. Graph views would graph all topics SI values against selected spatial and temporal hierarchy levels.
	Navigation	Navigation would allow the user to select different topical hierarchy schemes (if more than one) and then allow navigation through the hierarchy based on the selected view.

Table 2 Spatial views of sustainability indicators

Concepts	Structure	Detail
Organization	Predefined Geographic Units	Predefined geographical units structured in one or more spatially overlapping hierarchies with rules for aggregation within each hierarchy scheme.
	Scales	Data for each geographic unit would be reported for each one topical and temporal scale or be based on the aggregation rules for one or all scales (spatial, topical, temporal).
Schemes	GeoPolitical	Values summarized for a specific geopolitical spatial definition (city, county, etc.).
	Socio-Economic	Values summarized for a socio economic geography, census based (census tract block group), service area based (police beats).
	Ecosystem	Values summarized by spatial ecosystem units such biomes, watershed, climate zone.
Display	Data Display	Data could be viewed in a list, map, or graph view. In these views values for all spatial units at the selected spatial hierarchy level would be based on the selected temporal and topic scales. List views would list indicators based on a verbal description of the geography, such as city name, and display data values in numeric and or iconic formats. Map views would display a map, with data values displayed in map elements (chlorepeth, pie charts, etc). Graph views would be a graph of all geographic unit's SI values against selected topical and temporal hierarchy levels.
	Navigation	Navigation would allow the user to select different spatial hierarchies (if more than one is defined) and to browse different levels of the hierarchy by expanding or contracting the hierarchy list using the browsing tools for each view type.

1. Not all sustainability schemes include a topical hierarchical organization for the indicators. A lack of such a hierarchy even for SI schemes with only a few indicators will be difficult for decision makers to use when comparing several different alternative actions. Methods of organizing indicators in hierarchical structures need to be researched. (Gallopín 2004; Nyerges 2002)

2. Most indicators are not normalized, that is converted to some common scale that allows comparison of indicators created from different units of measure, as well as across topics, such as air quality and bio diversity. (Allard et al. 2004; Nyerges 2001) Though there is substantial literature on the normalization of indexes in general, there is little literature on application to sustainability indicators.

3. Most SI systems define indicators only at one or two spatial scales. This may be due to a lack of data at all geographic scales or a lack of methods to browse data at different scales (Leake et al. 2002). Emphasis should be placed in developing data and aggregation methods at various geographic scales.

Table 3 Temporal view of sustainability indicators

Concepts	Structures	Detail
Organization	Time line	This view would only be used when data is available for multiple time periods, which may vary by spatial and functional scales. Time data could be aggregated to longer time frames, such as a decade or century
	Scales	Data for each temporal unit would be reported for each topical and spatial unit for which data is available, or it would be based on the aggregation rules for one or all scales (spatial, topical, temporal)
Schemes	Snap Shot	Current, future, specific time (by year or month?)
	Delta	Change over time.
Display	Data Display	SI data would be displayed as list or timeline views. For snap shot data, this would simply be a reporting of the indicator for the date selected (past, present, or future). For change over time (delta) an actual numerical value of change could be displayed (magnitude of positive or negative change)
	Navigation	Navigation would allow the user to browse different levels of the hierarchy by expanding or contracting the hierarchy list using the browsing tools for each view type. Snap shot or change over time could be selected within each hierarchy level, except the lowest time level.

4. Most indicators do not provide rules for summation to higher topical and spatial levels. This is a critical aspect of the proposed framework in reducing complexity in areas of minor or no interest and expanding complexity in areas of high interest (Nyerges 2001).

5. There is general lack of historical indicator data most likely because indicators have only been widely reported for the last few years. Over time this will change but it will be important to maintain datasets of these indicators and maintain some consistency in how they are defined from one year to the next.

6. There is currently little mention in the literature on the use of sustainability indicators with futures analysis. This is may be in part because forecasting of the wide array of measures used in the sustainability indicators is currently not practical. However, as various environmental and urban systems models become more sophisticated, SI systems may be a practical way of simplifying results and scenario results could be viewed as a fourth hierarchical scope.

Conclusion

Decision makers usually have to deal with a wide variety of issues in which sustainability is only one factor. Though it is desirable for SI systems to become more detailed and complex so they can address this range of issues, decision makers will be looking for simple ways to understand sustainability. Thus there is a need for tools that can accommodate complex SI systems but have a high degree of flexibility providing both summary and details in a variety of levels of place and time. The proposed organizational and visualization framework presented here tries to provide flexibility reflecting the three major scales: topic, space, and time. This method provides a basis for a data standard that would allow a single user interface to display a wide range of indicator systems. More research and development of sustainability indicators will be needed to fully exploit this flexibility.

References

Allard F, Cherqui F, Wurtz E, Mora L (2004) A methodology to assess the sustainability of rehabilitation projects in urban buildings. Retrieved 9/15/05, 2005, from http://www.costc16.org/downloads/ WORKING%20GROUP%203/AllardF%20Sustainable%20Rehabilitation%20.pdf

Baker MP, Bushell C (1995) After the storm: considerations for information visualization. IEEE Computer Graphics and Applications 15(3): 12–15

Bots PWG, Twist MJWV, Duin RV (1999) Designing a power tool for policy analysts: dynamic actor network analysis. (Paper presented at the 32nd Hawaii International Conference on System Sciences, Hawaii)

Brandes U, Kenis P, Wagner D (2003) Communicating centrality in policy network drawings. IEEE Transactions On Visualization And Computer Graphics 9(2): 241–253

Central Texas Sustainability Indicators Project (2004) Fifth annual report. Central Texas Sustainability Indicators Project, Austin Texas

Chen C (2004). Information visualization: beyond the horizon: Springer, Berlin, Heidelberg, New York

Clark WC (2003) Sustainability science: challenges for the new millennium. University of East Anglia, Norwich UK (An address at the official opening of the Zuckerman Institute for Connective Environmental Research, September 4, 2003)

Consultative Group on Sustainable Development Indicators (2006). The dashboard of sustainability. Retrieved 3/17/2006, 2006, from http://www.iisd.org/cgsdi/dashboard.asp

Crampton JW (2001) Maps as social constructions: power, communication and visualization. Progress in Human Geography 25(2): 235–252

DiBattista G, Eades P, Tamassia R, Tollis IG (1994) Algorythms for drawing graphs: an annotated bibliography. Computational Geometry Theory and Applications 4: 235–282

EUROCITIES (2004) Towards a strategy on the urban environment: Eurocities statement: European Union

Forester J (1989) Planning in the face of power. University of California Press, Berkeley California

Gallopín G (2004) Sustainable development: Epistemological challenges to science and technology. (Workshop on Sustainable Development: Epistemological Challenges to Science and Technology, ECLAC.) ECLAC, Santiago de Chile

Gudmundsson H (2003) Making concepts matter: Sustainable mobility and indicator systems in transport policy. International Social Science Journal 55(2): 199–217

Hallisey EJ (2005) Cartographic visualization: An assessment and epistemological review. The Professional Geographer 57(3): 350–364

Jern M, Franzén J (2006, July 2006) "Geoanalytics": exploring spatio-temporal and multivariate data. (Paper presented at the Information Visualization 2006, London, United Kingdom)

Jolliet O, Friot D, Blanc I, Cenni F, Corbière-Nicollier T, Margni M (2003) Wp3 documentation, part 1: the sustainability model framework and indicators: The University of Minho, Portugal.

Klein G (1998) Sources of power: how people make decisions. The MIT Press, Cambridge Massachusetts

Lindblom CE (1995) Science of muddling through. In: McCool DC (ed.) Public policy theories, models, and concepts. Prentice Hall, Englewood Cliffs, New Jersey, pp. 142–157

Muller W, Schumann H (2003) Visualization methods for time-dependent data. (Paper presented at the 2003 Winter Simulation Conference)

Leake NL, Adamowicz WL, Boxall PC (2002) An examination of economic sustainability indicators in forest dependent communities in Canada: Sustainable Forest Management Network

Nardo M, Saisana M, Saltelli A, Tarantola S (2005) Tools for composite indicators building. European Commission, Directorate General, Joint Research Center, Ispra Italy

Nyerges T (2002) Visualizations in sustainability modeling: An approach using participatory GIS for decision support. Association of American Geographers, Los Angeles

Nyerges T L (2001, October 1–2, 2001) Research needs for participatory, geospatial decision support: linked representations for sustainability modeling. (Paper presented at the Intersections of Geospatial Information and Information Technology Workshop)

Prescott-Allen R (2001) The wellbeing of nations: a country-by-country index of quality of life and the environment. Island Press, Washington

Robertson GG, Mackinlay JD, Card SK (1991) Cone trees: animated 3d visualizations of hierarchical information. (Paper presented at the Conference on Human Factors in Computing Systems, New Orleans, Louisiana)

Shneiderman B (1992) Tree visualization with tree-maps: 2-d space-filling approach. ACM Transactions on Graphics 11(January): 92–99

SRP, S. R. P. (2004) Canada sustainability report: institute for Research and Innovation in Sustainability

Thomas J, Cook K (2005) Illuminating the path: the research and development agenda for visual analytics. Retrieved 11/1/2006, 2006, from http://nvac.pnl.gov/

Tufte ER (1990) Envisioning information. Graphics Press, Cheshire Connecticut

Tufte ER (1997) Visual explanations: Images and quantities, evidence and narrative. Graphics Press, Cheshire Connecticut

Venetoulis DJ, Chazan D, Gaudet C (2005) Ecological footprint of nations: Redefining Progress. www.redefiningprogress.org.

Wackernagel M, Onisto L, Linares AC, Falfán ISL, García JM, Guerrero AIS (1997) Ecological footprints of nations. Retrieved 9/15/05

Williams M (2004) Questvis and mdsteer: The visualization of high-dimensional environmental sustainability data. The University of British Columbia

Modeling Dynamic Land-Use Transition for Transportation Sustainability

Joonwon Joo

Introduction

Urbanization has increased significantly in the last century in order to accommodate drastic population growth worldwide. While it took 100 years from 1850 to 1950 for the human population to double in size from 1.2 billion to 2.4 billion, it has increased from 2.4 to approximately 6 billion in the 50 years since 1950 (Cohen 1995). During this 50-year period, dramatic urbanization has occurred around the world. To accommodate the population growth, cities have expanded and suburbs have been developed rapidly. This urbanization process has not only caused the loss of a significant amount of valuable agricultural land but has also increased various urban problems such as sprawl, air pollution, and traffic congestion. In addition, uneven development has worsened the economic and social problems of city residents.

The concept of sustainability emerged as an appropriate framework for abating the problems mentioned above. It was considered an ideal concept because it implied development in harmony with natural processes. The term sustainable development is widely used in planning and policy circles. The most well-known definition of sustainable development is "development that meets the needs of the present without compromising the ability of future generations to meet their own needs" (WCED 1987). However, the concept of sustainable development is still very broad and ambiguous, and it has been defined differently to meet the various purposes of individual research or policy.

Modeling of urban growth and change has often incorporated sustainable development as one of many objectives. However, the concept has often been focused on saving land supply and has seldom been applied to the complex interactions between built environment and human activities. Generally, urban modeling research focuses on projection of land use and transportation system changes in the future rather than

J. Joo
Multimodal Planning Division, Air Quality Policy Branch, Arizona Department of Transportation, 206 S. 17th Ave. 320B, Phoenix, AZ 85007, USA, E-mail: jjoo@azdot.gov

G. Steinebach et al. (eds.), *Visualizing Sustainable Planning*,   215
© Springer-Verlag Berlin Heidelberg 2009

suggesting possible policy alternatives for sustainable development linked to human activities. The primary objective of this research is to explore how sustainable development can be measured and projected in the context of land use-transportation interaction.

Research Design and Methods

To operationalize the concept of sustainable development, this research addresses one of the critical contextual elements, which is the relationship between land use and transportation. Sustainable development in this case would advance a land-use transition favorable for sustainable transportation activities such as more public transit use and more walking and bike riding. This concept of sustainable development is closely related to New Urbanism theory (Katz 1994). New Urbanists call for land-use patterns favorable for pedestrians and transit riders such as mixed-land use and highdensity residential areas. This new development paradigm is frequently addressed in the literature, but is not well understood. This research explores how the land use – transportation relationship can be modeled to address sustainable development.

Kevin Lynch in his 1981 book, *A Theory of Good City Form* discussed the idea of the city as an organism. If a city is an organism, then it is composed of cells that go through a cycle of birth, reproduction, and death. From this point of view, urban development, or urban growth, can be interpreted as the reproduction of cells in a living entity. Sustainable urban development may be interpreted as "evolutionary city growth in a positive way" (Lynch 1981). Several theories have been developed to operationalize the evolutionary paradigm of urban growth. This research breaks new ground by adopting Cellular Automata (CA) and Genetic Algorithms (GA) for investigating sustainable urban development based on land use-transportation interaction.

Several urban development simulation models with Cellular Automata theory have been developed (Barredo et al. 2003; Batty et al. 1999; Lie and Yeh 2000; Lobl and Toetzer 2003). However, the models do not project or suggest sustainable development, though many of them frequently address this issue. To address this problem, this research integrates Cellular Automata and Genetic Algorithms for the simulation models.

Most of the research on the link between urban form and travel behavior falls into three categories: simulation studies, aggregate analyses, and disaggregate analyses (Handy 1996). Each category of studies has its own limitations. The simulation studies do not empirically test the relationship between urban form and travel behavior (Handy 1996), whereas aggregate and disaggregate analyses have no power to test different urban development scenarios and they just investigate the current relationship. Addressing this methodological gap, the research suggests the integration of disaggregate, aggregate and simulation analyses and develops several empirical and simulation models.

The first model is developed to project urban form and land use transition with available accessibility and topography factors. It is a dynamic simulation model that predicts a future without any planning interventions. The second model is developed to investigate the current relationships among land-use patterns, accessibility, socio-economics, and travel behavior. It is a statistical model that empirically investigates the role of land use factors on transportation sustainability. The third model is developed to propose possible land-use patterns for sustainable transportation, and its impact on travel behavior and socio-economic factors. It is another dynamic simulation model like the first model, but is modified by the selected land use and site design strategies. It is developed on the basis of the results of the second model. In this paper, the results of the first model are presented.

To prove the validity of the proposed models, a case study will be conducted. The Phoenix metropolitan area in Arizona was selected for the case study because this area has experienced rapid urbanization with its dramatic population increase over the past several decades, and most land-use patterns are highly auto-dependent. New development has significantly increased traffic congestion and air pollution in the area. This area is also appropriate for exploring the relationship between transit and urban development since a new light rail system will be in operation in 2008.

Urban Growth Simulation

The urban growth model was developed to project the urban form changes of the Phoenix metropolitan area. The main purpose of this study is to consider how to achieve sustainable development. In this study, sustainable development is based on sustainable land-use patterns, density, and accessibility — Land Use Factors for Transportation Sustainability (LUFTS). To investigate LUFTS, the first task is to predict future urban growth and urban form changes, and then, in the projected urban form, LUFTS are predicted and suggested for sustainable development.

The urban growth model is developed by modified Cellular Automata (CA) theory. General CA models only consider the neighborhood effect, but the model suggested here estimates transition probability with suitability, accessibility, and neighborhood effect.

The model is calibrated to the 1995 Phoenix data by means of a trial and error approach to modify weight values of suitability, accessibility and neighborhood effect. The best method of model validation is to compare the simulation results with the actual city form. This study compares the projected 2000 Phoenix data with the actual 2000 Phoenix data. Many validation methods for simulation models have been developed, but evaluating models is difficult and is dependent upon the processes of interest (Turner, Wear, Flamm 1996). White et al. (1997) suggests three approaches: (1) a quantitative evaluation of the degree to which the land-use areas on the two maps coincide; (2) comparison of more abstract measures of the two maps, such as various measures of fractal dimension; and (3) a visual comparison

Table 1 Actual development status of the Phoenix area in 1995 and 2000

Year	Vacant	Developed	Non-developable	Total
1995	21,020	8,736	15,244	45,000
2000	20,332	10,241	14,427	45,000
+/− (%)	−688 (3.3)	+1,505 (17.2)	−817 (5.4)	0 (0.0)

The numbers of development status are cell counts

of the maps. They also admit that each of these has advantages and each has equal shortcomings.

To validate the model and find the appropriate weighting and mutation values, this study uses two quantitative validation methods — Kappa statistic and fractal dimension, and then confirms the results with visual comparison.

Table 1 shows the actual development status of the 1995 and 2000 Phoenix area. The numbers indicate cell counts from 250×180 grid system (500m $\times$ 500m for each cell). The change of cell counts shows how rapidly the Phoenix area has been developed. The developed area increased by 1505 (cells) during only a five-year period. Interestingly, the number of non-developable cells is quite different between 1995 and 2000. The reason for this discrepancy is that most of the cells were categorized differently between the two periods. Some non-developable 1995 cells became vacant cells by 2000. However, it hardly effects a five-year development simulation because most developed cells in 2000 were changed from the cells categorized as vacant in 1995.

As mentioned above, the development simulation is based on the weighted values of environmental suitability (k), accessibility (w_1) and neighborhood development status (w_2). Environmental suitability (0 or 1) is decided by mean elevation of each cell. Using raster DEM dataset, mean slope is calculated by GIS application. Comparing current elevations for developed and undeveloped areas (Jannerette and Wu 2001), k is decided ($k = 1$ if mean slope <3%, k = 0 otherwise).

Table 2 shows the simulation results and Kappa statistic for each scenario. Using cell-by-cell agreement, Kappa statistic compares simulation results and actual 2000 land-use data. It can be calculated by the following function:

Kappa = (Observed agreement − Chance agreement) / (1 − Chance agreement)

Kappa takes on a value of 1 with perfect agreement and 0 with perfect disagreement. The Kappa in the table shows that the model with weighted values of 10% accessibility and 90% neighborhood effect will project a future development pattern more accurately. Based on the Kappa statistic, accessibility explains only a small part of development and the goodness of fitness decreases when accessibility score increases. Although accessibility is not a major factor effecting urban development, the simulation result proves that it is somewhat related to the development pattern because the Kappa is higher when w_1 is equal to 0.1 or 0.2 than 0.

However, the Kappa statistic is not a perfect measure of similarity between maps because cell-by-cell agreement may not necessarily be the best standard for

Table 2 Urban form simulation results and Kappa statistic

w_1	w_2	Vacant	Developed	Non-developable	Total	Kappa Statistic
0	1	18,878	10,879	15,244	45,000	0.810
0.1	0.9	19,650	10,106	15,244	45,000	0.817
0.2	0.8	19,486	10,270	15,244	45,000	0.815
0.3	0.7	19,487	10,269	15,244	45,000	0.810
0.4	0.6	19,534	10,222	15,244	45,000	0.799
0.5	0.5	19,147	10,609	15,244	45,000	0.780
0.6	0.4	19,208	10,548	15,244	45,000	0.779
0.7	0.3	19,308	10,448	15,244	45,000	0.780
0.8	0.2	19,325	10,431	15,244	45,000	0.780
0.9	0.1	19,325	10,431	15,244	45,000	0.780
1	0	18,889	10,867	15,244	45,000	0.767

w_1 = weighting parameter for accessibility
w_2 = weighting parameter for neighborhood effect
The number of development status are cell counts

comparison. The most useful feature of the Kappa statistic (other than its objectivity) is in determining rank ordering across several maps (Monserud and Leemans 1992).

The next method for measuring goodness of fit is to compare fractal dimensions of actual and estimated development. Fractal dimension measure in this study is achieved by calculating the number of developed cells for each radius from the CBD area. Comparing with Kappa statistic, this method is more accurate for local measure of agreement. Figure 1 shows the area-radius relationship of actual and simulation development patterns of 2000 Phoenix area. Regression lines can be drawn for the area-radius relationship and β is a coefficient of each regression line. Table 3 compares β of estimated and actual development. The radius is limited within 28 km from the CBD because of grid cells' coverage.

Interestingly, unlike Kappa statistic, accessibility factor has more explanatory power in the fractal dimension for central area. The β of model with 0.5 of w_1, 0.5 of w_2 (216.70) is the closest to the actual 2000 dimension (217.17). According to this measure, about half of development can be explained by accessibility.

Finally, to confirm two comparison approaches above, the simulation results are compared visually. Many researchers agree that there is no perfect quantitative method to measure the similarity of two fractal forms. For complex fractal forms, visual comparison is often regarded as a more powerful tool than any quantitative technique (Mandelbrot 1983). Figure 2 compares actual the 1995 and 2000 Phoenix area with simulation results with various weight values (w_1: accessibility, w_2: neighborhood effect). As shown, the urban form projected from the model with 0.1 of w_1 and 0.9 of w_2 looks more similar to the actual 2000 Phoenix area than any others. The model with 0.5 of w_1 and 0.5 of w_2 (Fig. 2c) is the best-fit model for the second approach (fractal dimension), but its urban form is affected too much by the accessibility factor. Most new developments have occurred around the transportation network in this model.

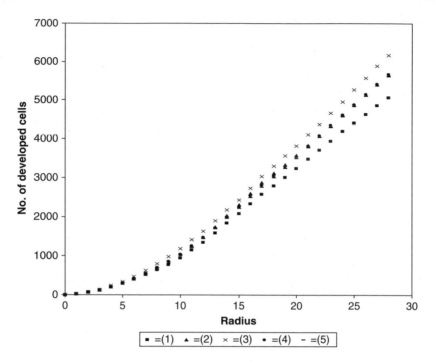

Fig. 1 Area-radius relationship of urban form

Table 3 Comparison of estimated and actual radial dimension of development

w_1	w_2	β^a	w_1	w_2	β
0	1	233.53	0.6	0.4	214.89
0.1	0.9	224.12	0.7	0.3	212.51
0.2	0.8	225.67	0.8	0.2	212.15
0.3	0.7	225.52	0.9	0.1	212.15
0.4	0.6	221.52	1	0	214.29
0.5	0.5	216.70			

[a]1995 Phoenix: 195.48, 2000 Phoenix: 217.17

Considering the results of the three comparison analyses, the model with 0.1 of w_1 and 0.9 of w_2 is selected to project future urban development of the Phoenix area. Because the mean slope of the developed area has increased, slope factor (k) is selected higher than the calibration (1995–2000) model ($k = 1$ if mean slope <5%, $k = 0$ otherwise). Population projection data is obtained from the Arizona State University GIS center and is used to adjust the simulation model for every 10-iteration. Table 4 shows the results of the simulation and how fast the Phoenix area will grow. Based on the simulation, developed area is doubled in 30 years.

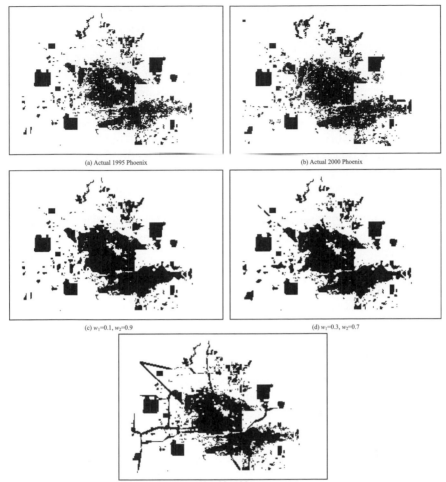

Fig. 2 Comparison of projected urban forms. (**a**) Actual 1995 Phoenix (**b**) Actual 2000 Phoenix (**c**) $w_1 = 0.1$, $w_2 = 0.9$ (**d**) $w_1 = 0.3$, $w_2 = 0.7$ (**e**) $w_1 = 0.5$, $w_2 = 0.5$

Table 4 Estimated future development of the Phoenix area

Year	Vacant	Developed	Non-developable	Total
2000	20,332	10,241	14,427	45,000
2010	16,327	14,246	14,427	45,000
2020	12,649	17,924	14,427	45,000
2030	9,610	20,963	14,427	45,000

The numbers of development status are cell counts

Simulated future urban forms are shown in Fig. 3. Comparing the urban form of the 2000 Phoenix area (Fig. 2a), new development is expected to occur mainly west of the Phoenix area. This is an expected result because the Indian reservation regions are located on the east and south of the area. The possibility of development in those regions is very low and the simulation model reflects this limitation. In fact, this expansion pattern reflects the current status of this area precisely because many new developments have recently occurred west of the city center.

The simulation results above indicate why sustainable development is necessary for a fast-growing area like Phoenix. Although it is difficult to suppress population growth, it is necessary to control development patterns in a sustainable way. At this point in the research (urban form projection), it is unclear how to control development patterns for sustainable development since, as of yet, sustainable development has only meant a reduction in the consumption of land supply (e.g., Yeh and Li 1998; Deal 2001; Barredo and Demicheli 2003). In order to address this issue, this study also estimates future land-use patterns for sustainable development.

Land-Use Change Simulation

Land-use change simulation is the second module for the future projection model. In the first module, future urban forms are projected above and now they serve as containers for land-use change simulations. Only CA (Cellular Automata) is used for urban form change simulation, but, within the projected urban forms, land-use changes are estimated by using modified CA and GA (Genetic Algorithms).

Table 5 compares actual land-use patterns of the 1995 and 2000 Phoenix area. There is a slight discrepancy in land-use code systems between the 1995 and 2000 land-use maps used in this study. Examining the source data, public land uses or non-developable areas are coded differently in many cases. As shown in the table, this discrepancy results in large increases in public land use and decreases in non-developable land use. However, it is not a significant factor for model calibration because CA functions for two land-use patterns are quite different with others. Public land-use cells come from mutations of other land uses and non-developable land-use cells are excluded from the function of the model. They do not influence other cells in the model run.

The simulation run from 1995 to 2000 is used to calibrate the model parameters, especially the mutation rate for each land-use category. Each newly developed cell can be mutated from one land-use category to another by a predecided mutation rate. This function compromises the deficiency of the CA mechanism and GA crossover function used in the model. As explained in the methodology chapter, only public land-use patterns are excluded from the CA mechanism in the model because public land uses such as parks and public facilities generally do not expand as other land-use patterns. In most cases they are developed after other land uses, especially residential land uses.

Fig. 3 Estimated future
urban forms of the Phoenix
area. (**a**) 2010 (**b**) 2020
(**c**) 2030

2010

2020

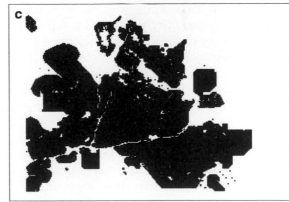

2030

Table 5 Actual land-use patterns of the 1995 and 2000 Phoenix area

Year	Vacant	Residential	Commercial	Industrial	Office	Public	Non-developable	Total
1995	21,020	5,461	285	505	206	2,279	15,244	45,000
2000	20,332	6,166	343	675	210	2,847	14,427	45,000
+/−	−688	705	58	170	4	568	−817	0
(%)	−3.3	12.9	20.4	33.7	1.9	24.9	−5.4	0.0

The numbers of development status are cell counts

Table 6 Land-use change simulation results and Kappa statistic

Year 2000	Vacant	Residential	Commercial	Industrial	Office	Public	Non-developable	Total	Kappa
Simulation	19650	6251	350	598	265	2642	15244	45000	0.801
Phoenix	20332	6166	343	675	210	2847	14427	45000	–
+/−	−682	85	7	−77	55	−205	817	0	–
(%)	-3.5	1.4	2.0	−12.9	20.8	−7.8	5.4	0.0	–

The numbers of development status are cell counts

Table 6 shows the comparison of simulation results and the actual 2000 Phoenix area. The Kappa statistic is relatively high (0.801) in this simulation. Generally, if Kappa is greater than 0.7, the degree of agreement is very good (Monserud and Leemans 1992).

As analyzed for urban form simulation, the area-radius relationship is also examined to determine the accuracy of the model (Fig. 4). Within 28 km from CBD, Overall land-use patterns of simulation are similar to the actual city. The radial dimension of each land-use pattern is also very similar between the two (see Table 7). However, as shown in the graphs in Fig. 4, the cell distributions of office and public land uses in the inner city are quite different between real city and simulation. The reasons for these results are different for office and public uses. For office land use, CA mechanism (neighborhood effect) creates this difference. Around the CBD, office land uses are concentrated, so there is a good possibility that the surrounding vacant cells will become office land uses as indicated by simulations controlled by CA used for the model. For public land use, the discrepancy of source data creates this difference.

Comparing simulation and actual city maps for public land uses (Fig. 5e), there is a relatively large public land use area in the center of the actual city (right map) that does not appear in the simulation (left map). In fact, it is not a new public area developed between 1995 and 2000. This difference comes from one of the discrepancies in land-use boundaries between the 1995 and 2000 map data. This public area on the 2000 map already existed in 1995, but when it was divided for grid cells for simulation, the grid cells were not assigned to public land use cells on the 1995 map.

Visual comparison of each land use is also very similar between simulation and real city (see Fig. 5). As explained above, office land uses in simulation map (d) are concentrated around CBD. Interestingly, commercial and industrial land uses are developed around major road networks in Phoenix, so simulation also follows this

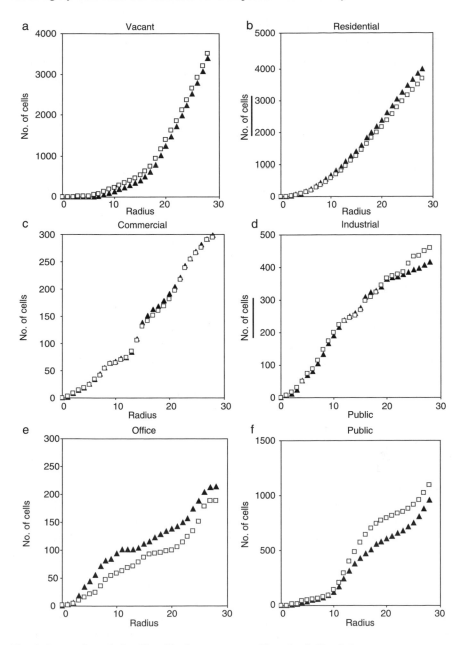

Fig. 4 Area-radius relationship of land-use pattern. □ Phoenix, ▲ Simulation

pattern of development. The land-use maps in Fig. 5 show the different development patterns of each land use. Residential and industrial land uses follow an agglomeration economy principle while commercial and office land uses are inclined to separate with each other. These phenomena tell how simulation models should be developed when much iteration is required.

Table 7 Comparison of estimated and actual radial dimensions of land-use patterns

Year 2000	Vacant	Residential	β Commercial	Industrial	Office	Public
Simulation	112.66	153.25	11.45	16.37	7.14	35.92
Phoenix	117.82	140.50	11.20	17.36	6.33	44.03

The numbers of development status are cell counts

Using the parameters adjusted by the 1995–2000 simulation model, future developments are projected. The simulation results are summarized in Table 8. The results show that land supply will be reduced quickly due to a rapid increase in population by 2030. However, the development phase is gradually slowed as shown in the table. This happens because some of new development may take place outside of the study site (the grid system). The grid system covers most of Maricopa County where the Phoenix metropolitan area is located, but new development has recently expanded to surrounding counties.

The rate of office land use development is quickly lowered after 2020 because most of available lands in the central area, which is favorable for office uses, are depleted by that time.

The land-use maps generated by simulation results show how land-use patterns in the Phoenix area will change over the next 30 years (Fig. 6). As new developments occur on the periphery, large commercial developments follow because these newly developed areas are far from the current commercial areas of central Phoenix. New residential areas will be developed far from the currently developed areas, but, as mentioned earlier, north, east and south of the site are non-developable areas such as high-slope conditions and Indian reservations. For this reason, new residential areas after 2020 are expected to occur northwest and southwest of the site. The southwest area is especially developed with new commercial, industrial and office land uses as well as residential land use, since this area surrounds interstate-highway 10 (I–10).

As expected, future land-use patterns simulated by the model do not demonstrate transportation sustainability as defined by this study. In addition to low-density development (sprawl), mixed land uses are not apparent and development occurs around major road networks rather than public transportation systems. These future development patterns are expected if they are extrapolated through historic data. However, if the model parameters include sustainability indicators, different, more sustainable scenarios may be possible.

Discussion and Further Research

The automobile became the transport technology that shaped the city around the Second World War. It made it possible to develop in any direction, and the city began to decentralize and disperse (Newman and Kenworthy 1996). Low-density

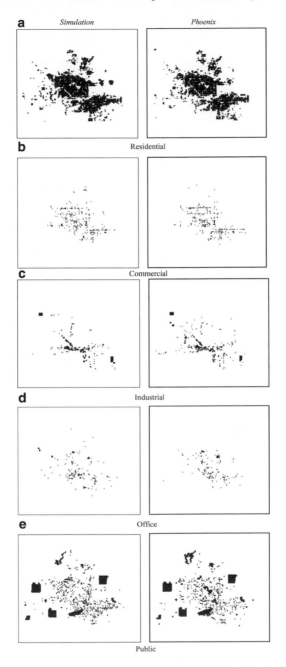

Fig. 5 Comparison of projected and actual land-use patterns (**a**) Residential (**b**) Commercial (**c**) Industrial (**d**) Office (**e**) Public

Table 8 Land-use pattern changes of the Phoenix area

Year	Vacant	Residential	Commercial	Industrial	Office	Public	Non-developable	Total
2000	20332	6166	343	675	210	2847	14427	45000
2010	16327	8550	506	936	302	3952	14427	45000
+/−	−4005	2384	163	261	92	1105	0	
(%)	(−19.7)	(38.7)	(47.5)	(38.7)	(43.8)	(38.8)	(0.0)	
2020	12649	10665	785	1144	384	4946	14427	45000
+/−	-3678	2115	279	208	82	994	0	
(%)	(−22.5)	(24.7)	(55.1)	(22.2)	(27.2)	(25.2)	(0.0)	
2030	9610	12433	1040	1282	417	5791	14427	45000
+/−	−3039	1768	255	138	33	845	0	
(%)	(−24.0)	(16.6)	(32.5)	(12.1)	(8.6)	(17.1)	(0.0)	

The numbers of development status are cell counts

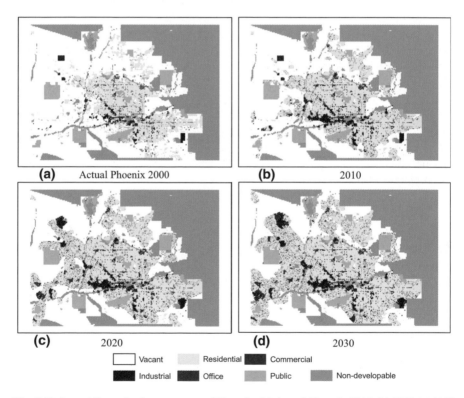

Fig. 6 Estimated future land-use patterns of Phoenix. (**a**) Actual Phoenix 2000 (**b**) 2010 (**c**) 2020 (**d**) 2030

housing became more feasible, and, as a reaction to the industrial city, town planning began separating functions by zoning. Cities such as Phoenix, Denver, and Houston were developed in the automobile era.

However, over the last several decades there has been an increased concern in metropolitan regions with the decline in air quality, increased congestion in both urban and suburban areas, and the negative impacts on the natural environment resulting from land development patterns overwhelmingly favorable to the automobile. In order to address these concerns and meet national air quality standards more recent emphasis has been placed on encouraging the development linked to public transportation systems (Badoe and Miller 2000).

The research suggests specific land use patterns for sustainable travel behavior. In this research, Sustainable travel behavior simply means increasing transit use, walking and bike riding, and decreasing private auto use. This can be a critical foundation for increasing economic, environmental, and social sustainability, which are the main objectives of sustainable development. Sustainable land use and travel behavior decreases energy use and traffic congestion, and preserves the environment by reducing air pollution. A pedestrian – friendly land-use pattern renews neighborhood perception because it increases the chance to meet neighbors on the street.

As mentioned above, this paper presents only the first model – future projection model. Based on this model, a statistical model will be developed for investigating current relationships between land use and travel behavior, and the second simulation model will be developed for suggesting sustainable development in the context of land use – transportation interaction.

A complete generic model developed here may be applied to any place where high auto dependency is one of the critical issues in the path of sustainable development. By using location-specific data for the suggested model in this research, researchers and practitioners can not only project future urban development patterns but also consider alternate development patterns connected to sustainable development and transportation sustainability.

References

Badoe DA, Miller EJ (2000) Transportation-land-use interaction: empirical findings in North America, and their implications for modeling. Transportation Research Part D: Transport and Environment 5:235–263

Barredo JI, Kasanko M, McCormick N, Lavalle C (2003) Modelling dynamic spatial processes: simulation of urban future scenarios through cellular automata. Landscape and Urban Planning 64:145–160

Barredo JI, Demicheli L (2003) Urban sustainability in developing countries' megacities: modelling and predicting future urban growth in Lagos. Cities 20:297–310

Batty M, Xie Y, Sun Z (1999) Modeling urban dynamics through GIS-based cellular automata. Computers, Environment and Urban Systems 29:381–393

Deal, B (2001) Ecological urban dynamics: the convergence spatial modeling and sustainability. J Building Research and Information 29:381–391

Handy S (1996) Methodologies for exploring the link between urban form and travel behavior. Transportation Research Part D: Transport and Environment 1:151–165

Jenerette GD, Wu J (2001) Analysis and simulation of land-use change in the central Arizona-Phoenix region. Landscape Ecology 16:611–626

Katz P, Scully VJ, Bressi TW (1994) The new urbanism: toward an architecture of community. McGraw-Hill, New York

Li X, Yeh AGO (2000) Modelling sustainable urban development by the integration of constrained cellular automata and GIS. Int J Geographical Information Science 14:131–152

Loibl W, Toetzer T (2003) Modelling growth and densification processes in suburban regions: simulation of landscape transition with spatial agents. Environmental Modelling and Software 18:553–563

Lynch K (1981) A theory of good city form. MIT Press, Cambridge Massachusetts

Mandelbrot B (1993) The fractal geometry of nature. In Nina Hall (ed.) New scientist guide to chaos. Penguin, Harmondsworthpp 122–135

Monserud RA, Leemans R (1992) Comparing global vegetation maps with the Kappa statistic. Ecol Modell 62:275–293

Newman P, Kenworthy J, Vintila P (1992) Housing, transport and urban form. Background Paper 15, National Housing Strategy Dept of Health, Housing and Community Services, Canberra, Commonwealth of Australia.

Turner MG, Wear DN, Flamm RO (1996) Land ownership and land-cover change in the southern Appalachian highlands and the Olympic peninsula. Ecol Appl 6:1150–1172

Yeh AGO, Li X (1998) Sustainable land development model for rapid growth areas using GIS. *Int J Geographical Information Science* 12:169–189

White R, Engelen G, Uljee I (1997) The use of constrained cellular automata for high-resolution modelling of urban land-use dynamics. Environment and Planning B: Planning and Design 24:323–343

World Commission on Environment and Development (WCED) (1987) Our common future. Oxford University Press, Oxford

Sustainable Phoenix: Lessons from the Dutch Model

Jesus J. Lara

Abstract In only fifty years, the Phoenix metropolitan area has expanded from a small desert town into one of the largest urban areas in the United States. Today, it has one of the fastest rates of growth in the nation with an annual rate of 4.5%. This area has grown during a period in urban development that largely ignored local topography, climate, culture, and history. The result has been a sprawling metropolitan area with an ever increasing ecological footprint and a standardized urban design and infrastructure that works against its environmental setting rather than with it. Currently, the city of Phoenix is going through a process of urban revitalization with an increasing demand for urban living and commerce. This research explores sustainable urban design and its potential applications in the metropolitan Phoenix area through an investigation of the Dutch model. The Dutch have successfully dealt with sustainable urban design approaches and their practices represent an unusual learning opportunity for Phoenix. The Netherlands' experience suggests three strategies/themes for rendering Phoenix a more sustainable urban form. These include the strategic planning and development of urban extensions, compact infill, and modernizing infrastructure.

Problem Statement

Like many urban areas in the U.S., the Phoenix urban area reflects development that has been driven by a powerful economic engine. This economic force has produced uncontrolled development that is not balanced by social, historical, aesthetics, or environmental concerns. As in many areas, suburban developments have replaced the rural landscape with generic sprawling subdivisions, retail boxes, six lane freeways, and large expanses of parking lots; mini-malls, billboards, and fast-food

J.J. Lara
Austin E. Knowlton School of Architecture, The Ohio State University, 291 Knowlton Hall, 275 West Woodruff Avenue, Columbus, OH 43210-1138, USA, E-mail: lara.13@osu.edu

G. Steinebach et al. (eds.), *Visualizing Sustainable Planning*,   231
© Springer-Verlag Berlin Heidelberg 2009

restaurants sit next to historic buildings, diminishing the sense of place and urban character. All these elements and planning practices have contributed to the degradation of urban design and the quality of life in our city.

In the quest to find ways to improve the urban condition, getting the design and quality of the urban fabric right are crucial conditions for creating more sustainable communities. The right quality of urban fabric means that we must create well-designed places that put people first and make efficient use of the available space and environmental resources. Well-designed places require critical and multifaceted policy, analysis, and design, taking into account the land, history, society and economics. In addition, well-designed places are urban interventions that have to be able to respond to current forces that make it difficult to achieve high-quality design such as population migration and growth and rapid urbanization. In order to address these complex issues, we must search new ideas and fresh thinking that can restore and improve degraded communities, rather than utilizing old and obsolete formulas to design our urban areas.

Research Objectives

The purpose of this research is to examine current sustainable urban design approaches and strategies in the Randstad region of the Netherlands. Very little is currently known about the transfer of knowledge through the exploration of best practices in the fields of planning and urban design. The primary body of knowledge in planning and urban design is contained in the written and visual documentation of case studies (Francis 2003). According to some authors, case studies serve as the collective record of the advancement and development of knowledge in urban design (Coupland 1997; Beatley 2000; Francis 2003).

Throughout this research process I have sought to answer questions in one major area of sustainable urban design: what are the best practices of sustainable urban design in the Netherlands and are these practices transferable and adaptable to the Phoenix area? Through interdisciplinary cooperation and through both design initiative and the creation and implementation of policies that support compact cities, the Dutch have made impressive advances in the direction of more sustainable futures in urban livability.

Issues of Transferability

While the intentional spread of case study areas over the Dutch geographical region will confirm the notion that sustainable urban design has undeniable international relevance, the transferability of good ideas between both continents remains problematic. No two places are the same, therefore, it is important to note the significant differences in the political, economic, geographic, and cultural systems between

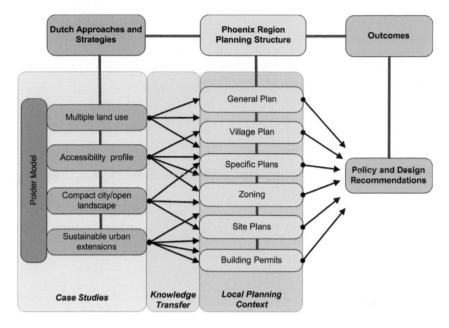

Fig. 1 Overall research approach. The idea here is to identify gaps or places within the planning structure of Phoenix, where the transfer of knowledge of specific approaches and strategies from the Dutch might occur given consideration of the general local context

urban areas in the Randstad in the Netherlands and the Phoenix metropolitan area. The goal of this research was not to duplicate or to import what the Dutch are doing in terms of achieving sustainable urban design. The goal is to transfer the knowledge from Dutch sustainable urban design approaches and provide suggestions for how these adaptations might occur in the context of Phoenix (see Figure 1).

Learning from Dutch design and planning practices is not new. According to Tim Beatley (Beatley 2000) there is already a rich history of powerful environmental planning and sustainability ideas firmly established here that originated in the Netherlands. Traffic-calming techniques and methods that are becoming increasingly popular in the United States were first applied in the Netherlands. The concept of *woonerf* (living street) has contributed to the enhancement of urban livability. Car-sharing, an increasingly important European practice is now beginning to find application here in the United States.

Methodology

The selection of the case study areas reflects a desire to identify the edge of innovation in sustainable urban design oriented projects. Based on the case studies, this research presents a model of performance evaluation for existing urban conditions in Phoenix and draws recommendations for further developments that result in concepts for new projects. Successful sustainable urban design initiatives in Phoenix will need to be developed specifically from local resources and in response to local conditions—and it will be shown that this is exactly the approach that proved critical to bringing about the "innovative edge" perceptible in the case studies discussed here.

Selected Dutch case studies include the following:

1. Urban extensions: VINEX locations: Ypenburg (Den Haag), Ijburg (Amsterdam), Leidsche Rijn (Utrecht).
2. Compact infill: Den Haag New Centrum, New Ceramic (Maastricht)
3. Modernize infrastructure/retrofitting urban areas: Kop Van Zuid (Rotterdam), Eastern Harbors (Amsterdam).

Practical Innovations in the Netherlands: The Dutch Model of Urbanization

With the increase of globalization in almost every aspect of our lives, it is critical that we take a closer look at how others are approaching similar issues of growth and urbanization. Planning and design practices in the Phoenix metropolitan area can draw inspiration from the way other metropolitan areas address similar issues associated with growth and urbanization. The Randstad region in the Netherlands is an excellent place from which to learn.

What can the Phoenix area learn from planning and design practices in the Netherlands? This research sought answers to this question by examining urban design and planning practices in the Randstad. The institutional contexts, planning traditions, and approaches of the study area are obviously different from Phoenix. Although at first glance one can say that both study regions are totally different, yet if we look deeper into the current challenges facing the Netherlands today, it becomes apparent that they are similar to those challenges facing the Phoenix area. For instance, some of the problems affecting both regions include urban pressures such as population growth, urban decline, preservation of open space, and containment of sprawl.

Current Planning Support in the Netherlands

In the case of the Randstad, trends and policies from various levels of government influenced the development of the region. An important contribution to the successful implementation of sustainable design practices in the Randstad, and in the Netherlands in general, has been strong support from the national government and the Ministry of Housing and the Environment (VROM). Since the mid-1980s and the introduction of the Fourth National Planning Report (1988), issues of sustainability have had high priority on the government's agenda. Over the past 25 years, the Ministry of Planning has had significant control over urban redevelopment initiatives in cities and the implementation of new infrastructure.

Key Factors that Contribute to Successful Planning and Design

Achieving sustainable development requires the attention and participation of all parties involved in the process, beginning with the initial stages of design and implementation, and ending with post-occupancy project evaluation. In addition to participation, government support is a critical factor to the success of any sustainability initiative. Having support from the government through policies and regulations can further ensure the realization of any sustainability initiative (Beatley 2000, 2004; Boonstra 2001). In the Netherlands, the success of sustainable initiatives can be attributed to Dutch sustainable building and planning policies. These policies can be explained by a combination of geographic, climatic, and perhaps, more importantly, cultural factors. Some of the key factors that have facilitated sustainable development in the Netherlands include, but are not limited to

1. a strong national government with embedded powers at different levels of the national planning system. The national government exerts power over important infrastructure, such as freeway, rail systems and airports;
2. stipulations in national planning reports that support sustainable development, such as the case of the 1993 National Spatial Policy Plan;
3. a general awareness about the existing expansion conflicts among urban areas in the Randstad and the need for preservation and improvement of the Green Heart (i.e., the central part of the Randstad where the primary activities are characterized by agriculture, nature and recreation), and
4. lastly, the budget for sustainable development practices originates from the national tax base, and not from the local tax base.

Dutch Approaches: Shifting Emphasis

In addition to the influence of the factors mentioned above, in the Netherlands there has been a shift in the way in which professionals as well as the general population approach environmental issues. For instance, with the influence and support of the national government during the last 25 years, there has been a shift in the way planners, designers, and the general population approach and implement sustainable practices (Bueren 2001; Vlassenrood 2004). For instance, there has been a shift from simply the use of sustainable materials to the implementation of ecological themes in a larger context. Some of these themes include but are not limited to water, traffic and transport, green structures, and cultural heritage, which have become important themes in sustainable design. Also, the sustainability emphasis has shifted from the construction of a single building block toward much larger building scales such as neighborhoods, districts, and entire communities, which tends to have a larger impact on urban sustainability.

With the increase in urban densities and the lack of available space for new development, and in order to promote urban living and increase the vitality of cities, attention has shifted during the last two decades to existing leftover urban space and the recycling of brownfield areas for new redevelopment. Consequently, the focus has shifted from building new housing and offices in green fields (undeveloped land at the urban fringe), to the reuse of existing buildings.

In addition, there has been a shift from the implementation of physical systems to a focus on social systems. The social environment in which sustainable building takes place has become important too. There is no ecological sustainability without social sustainability. Issues such as social segregation, quality of life, and community involvement have acquired a prominent place on the agenda of Dutch politicians and designers.

Urban Conditions in the Phoenix Metropolitan Region

Phoenix is no stranger to rapid growth and to the problems that come with it. Over the last century Phoenix has persuaded prosperity through urban expansion. The Phoenix region is so new that almost 90 percent of its infrastructure has been built within the last 50 years. Most of the development in the region can be attributed to three kinds of growth: land area, population, and infrastructure.

Land Area Growth

Since the 1950s, metropolitan Phoenix has been expanding rapidly and losing desert and agricultural areas to urban uses. Between 1975 and 1995, the metropolitan Phoenix urban area more than doubled (Policy 2000). Metropolitan Phoenix is contained within Maricopa County, and consists of 24 cities and towns that make up the Maricopa Association of Governments (MAG). Urban development now covers more than 40 percent of the MAG planning area. The area is 9,226 square miles, which is larger than New Jersey and other four states. The city of Phoenix is larger than Paris, Rome, San Francisco, and Manhattan combined; it can take two hours to go from one end to the other (Lincoln Institute of Land Policy 2003).

In Phoenix, urban land area grew from 117 square miles in the 1950s to 1,760 square miles in 2003. The urban region has grown an average of 47 percent each decade since 1960 (Governments-MAG 2005). In addition, between 2000 and 2004, the urban edge expanded by 55,000 acres – approximately 33 acres a day. It is estimated that the growth of Phoenix's urban footprint is equivalent to a half a mile each year (Governments-MAG 2005). As a result of this rapid growth the region is experiencing impacts such as increasing urban heat island effect due to the excessive amount of paving, reduced accessibility to open space, fragmentation of natural vegetation, and an increasing loss of native species such as the creosote bush (*Larrea tridentate*). Most of the urban edge land lost to development has been sold to homebuyers who want to live close to nature in a rural environment and yet retain all the amenities and services that cities have to offer.

Population Growth

Population growth in Phoenix has contributed to the rapid development in the region. During the past few decades, Phoenix has become one of the fastest growing and most rapidly diversifying metropolitan regions in the United States, with an annual average population growth rate of 4.5%. The annual population growth in the area is almost three times the national average (Downs 2001). In 1950, the region had 100,000 people; by 2003 the region reached 3.3 million. It is expected that the population will reach 7 million by the year 2025.

With the addition of at least 100,000 people every year, growth in the Phoenix area has been described as hyper-growth. Over the last 30 years, an average of 127 new residents moved to the Phoenix area every day (Morrison Institute for Public Policy 2000). Most of the growth in the metropolitan area took place between 1970 and 1998. From just 1990 to 1998, the population increased 31 percent, thanks in large part to the arrival of an average of 57,000 new residents a year (Governments-MAG 2005). About the same time the area became one of the 25 largest metropolitan areas in the nation.

Infrastructure Growth

In addition to land area and population growth, infrastructure growth has had a great impact in the development of the region. Aside from the many benefits to the region's economy, there are also challenges to its infrastructure and natural resources, particularly in the urban fringe. Some of these challenges include more traffic, longer commutes, and air pollution, all of which result from a dramatic influx of drivers and homebuyers. Research done by the Morrison Institute (2001) shows that most of the communities where the new growth is taking place are relatively small, new, and inexperienced when it comes to dealing with issues associated with the control and management of rapid growth. Therefore, overwhelmed by rapid growth, these inexperienced communities find themselves unable to cope with the problems associated with rapid development. For instance, when growth outpaces the rate at which tax rolls increase, cities can not provide all the services and infrastructure that the new development demands. This situation increases the potential for issues to develop that affect the region as a whole.

Practical Innovations in the Netherlands: Case Studies Uncovering Design and Policy Principles

The case studies considered here enlarged upon the information gained from the literature review. The three different types of cases were selected to represent some of the most innovative strategies in planning and design in the Netherlands.

The Randstad region in the Netherlands is an excellent example because it is home to some of the greatest concentrations of dynamic and innovative actions in the world.

The projects documented in these seven case studies are organized under three main categories (1) urban extensions, (2) compact infill, and (3) retrofitting urban areas. These case studies demonstrate urban design and planning principles that reflect the desire to identify the "edge" of innovation in sustainable urban design oriented projects. The diversity of project types is intentional, covering a range of activities, scales, and locations across the Randstad to ensure wide appeal to the general public, and to illustrate as many principles as possible. Selected projects demonstrate the practical application of urban design principles, the benefits that come from good practice, and the areas where further improvements could be made. The methods for each case study analysis consisted of collecting background information for each site including summary facts and project statistics, a description of the design process, urban design issues, an evaluation of the project's successes and limitations, lessons learned, and the value gained.

Urban Extensions/VINEX Locations

In the Netherlands the quality of urban design and the conditions for urban planning have been determined for the VINEX sites by the Ministry for Housing, Spatial Planning and Environment (VROM). In qualitative terms, the VINEX projects represent an opportunity to design genuine new city areas with an urban character, with compact building development, varied programs, and a strong relationship with the existing city's built environment and social connections. In some cases these opportunities for relationship are achieved through a combination of public transport and business premises or through the creation of socio-cultural facilities. The three VINEX projects explored in this research include (1) Ypenburg (Den Haag), (2) Ijburg, (Amsterdam), and (3) Leidsche Rijn (Utrecht).

The VINEX policy is an urban compaction policy designed to meet regional needs for housing, employment, health care, and urban facilities on a regional scale. The VINEX projects were introduced in the Fourth Policy Document on Spatial Planning Extra (VINEX 1993), and are supported by policies that promote the implementation of compact cities. This policy contained massive house-building targets. Mobility objectives played a key role in the policy. As a result, the location, layout, and accessibility of the new VINEX housing developments were designed to help reduce non-essential car use. This concentration of urban development aims to provide the critical mass for urban services and amenities, limit further urban expansion and encroachment in rural areas, and curb the growth in mobility (Ministry of Housing 2000).

Compact Infill

Compact infill development is an urban strategy frequently implemented in the Netherlands. It is intrinsically connected to the national compact city policy, which has been highly regarded by the planning authorities of the Netherlands since the 1980s. Given its small size and high density, cities in the Netherlands are the most popular places to live and work. The compact city is seen as the answer to two main problems that the economic center of the Netherlands, the Randstad, as well as other more densely populated areas of the country were facing: (1) fast urbanization of open space, and (2) the continued increase of mobility. The Ministry of Housing and the Environment (VROM) introduced the compact city policy as part of the Fourth Physical Planning Report Extra in 1990 in order to address these two urban issues. The idea was that compact city polices would contribute to the spatial quality of both urban and rural areas. In addition, one of the expected positive byproducts of compact city policies was a positive effect on the environment because it would reduce traveling distances.

The next two case studies under this category include two major projects dealing with compact infill development in highly urbanized areas: (1) *Den Haag New Centrum* in the city of Den Haag, and (2) *Maastricht: The New Ceramique* in the city of Maastricht. Although the latter case study is not located within the Randstad region, it represents a very interesting and successful example of a compact urban infill project.

Modernizing Infrastructure/Retrofitting Urban Areas

Space in the Netherlands is scarce, so any available space is highly valued. Therefore, the sustainable use of any available space is a major concern of the Dutch planning authorities. The increasing concentration within urban areas has forced planning authorities to provide alternative spaces for redevelopment within existing urban areas. As a result, there has been a shift away in the way the Dutch approach land development. For instance, within the last decade there has been a shift from newly built dwellings and offices to the adaptation of existing buildings for new uses.

Some of these new developments are taking place on former brownfield areas, leftover spaces from planning, and in former industrial areas. In these situations, the redevelopment of structures with obsolete uses and functions represent a feasible solution for development within the urban context. In this way, the Randstad has been experiencing the urban transformation of inner-city industrial areas, harbors, factories, public utilities, and hospitals over the past ten to twenty years. These new urban spaces play an important role in adapting the city to the changing demands of its residents.

Under this category I explored two examples of urban intensification through the process of adaptation and retrofitting of former industrial areas including (1) Kop van Zuid in Rotterdam, and (2) Eastern Docklands-Java Island in Amsterdam.

Fig. 2 Broad comparison of both study areas: (**A** & **B**) In the Randstad compact city policies are mandatory and are supported at all government levels. (**C** & **D**) In Phoenix multiple land use policies are nonexistent that results in abandoned central city, due to obsolete planning and zoning regulations. In Phoenix they are still building the suburban shopping malls from the 70s and 80s that basically isolate residential areas only accessible by car diminishing the urban life

Phoenix: Conditions, Assets, and Sustainability

Currently, the city of Phoenix is going through a process of urban renaissance with an increase in demand for urban living and commerce. Demands for a new central city include many of the integral components of sustainable urban design, including pedestrians, housing diversity, live-work environments, arts and entertainment, efficient public transportation systems, institutional assets (government, universities, and the medical industry), and a renewed interest in "sense of place" in daily living. It is said that Phoenix has most of the major building blocks to become the heart of a new regional economy. However, to keep up with the demands of this urban renaissance, and at the same time provide the required physical and social infrastructure for its residents, it is crucial for the city and its surrounding metropolitan area to learn from best practices elsewhere.

Phoenix Assets

Critics, scholars, and policy makers agree that the Phoenix metropolitan area is ready to take a strategic approach to urbanization. This positive view is the result of the economic growth in the region during the last ten years, and most recently, of an increase in the science and technology sector. Experts agree that for the Phoenix metropolitan region to continue to prosper, three overlapping spheres of influence must be considered: the creation of knowledge capital, the enhancement of social capital, and the preservation of natural capital (Lincoln Institute of Land Policy 2003). Elements of these three overlapping spheres are already present in the many assets that the Phoenix metropolitan area possesses. Some of the many assets of Phoenix include, but are not limited to

- Population: Metropolitan Phoenix's population, currently 3.5 million, is one of the fastest growing and diverse in the country. With nearly 5 million more people expected in the next 25 years, smart planning will be crucial to prevent boom from becoming bust (Morrison Institute for Public Policy 2006).
- Science and Technology: The biotech cluster in the middle of downtown is booming. In March, the city opened the Phoenix Biomedical Center, a 28-acre campus that houses both the Translational Genomics Research Institute and the International Genomics Consortium (Morrison Institute for Public Policy 2006).
- Tourism: Phoenix is an attractive destination to newcomers because of its climate, its access to recreational areas, and its location within the nation.
- Economic growth: Population and economic growth has been increasing steadily in the region. Experts agree that metropolitan Phoenix is still experiencing phenomenal growth, with nearly 700,000 new residents (31 percent increase) and approximately 500,000 more jobs in less than 10 years (Morrison Institute for Public Policy 2000). A reason for this growth has been substantially lower home prices compared to most other large metropolitan areas. Lower home prices attract families seeking to own homes, as well as real estate investors, home-builders, construction workers, and all the support services.

Although the Phoenix metropolitan area has the potential to become a major economic center in the region, it is important to ensure that all the region's assets are effectively managed to provide the optimum conditions for a more sustainable future. In the same manner, the area needs to be able to face the challenges that growth presents, such as undertaking its first light-rail project, scheduled to open in 2008; rising real estate costs and other effects of a strong, long-term population increase; the largest population increase in the western states with more than 2,000 residents per week; and the needs of a population that is expected to double in the next twenty years.

Recommendations for Phoenix

The Phoenix metropolitan area is expected to grow by seven million people over the next twenty years. Without changes in development policy and practice, this growth will continue to be characterized by low-density urban sprawl. Learning from the Dutch best practices can provide a more sustainable edge to current development practices in Phoenix. However, for Phoenix to adapt some of the approaches and strategies from the Dutch system (compact city policies, multiple land uses, and the adaptive reuse of obsolete urban infrastructure), it is crucial to improve in three critical areas: Phoenix should (1) *apply a holistic approach to policy and design*; (2) *provide policy and design that reflect local context*; and (3) *invest in the human component of policy and design.*

- The first recommendation is to approach urban issues in a more holistic manner. To achieve a sustainable urban design through the transfer of knowledge from the Dutch system, it is crucial to comprehend the interconnectedness that exists among different aspects of the design and policy process. An understanding of the interconnectedness of the elements of the built environment can be attained through a more holistic approach to design at both local and regional levels. A holistic approach does not require that those involved in the design and planning process know and understand everything that they are dealing with, but rather that they acknowledge what they know and understand what they do not (Mclennan 2004). In other words, a holistic approach involves collaboration with those who might hold parts of the solution that are crucial to planning and design processes, and can help to seek the linkages to the overall vision. A holistic approach can help to widen the circle of comprehension by understanding the connections that exist between all the aspects involved in the design of the built environment.
- The second recommendation is to understand its context and provide policy and design interventions that reflect its locality. Planners and designers in the U.S. have tended to neglect and underestimate the importance of designing for place, and the role that locality plays in creating more sustainable urban environments. In Phoenix, climate and place are critical factors in planning and design. Sustainable urban design in this region should enhance the quality of life, and recognize and support people's evolving sense of well-being. Well-being includes a sense of belonging, a connection with nature, and respect for the ecological integrity of natural systems (Corbett and Corbett 2000). In Phoenix, the sense of belonging can be accomplished through the implementation of sustainable urban design guidelines that provide identity. The design guidelines would be based on local socio-cultural and natural environments, and would provide the necessary conditions for a more sustainable urban desert environment. The objective of these guidelines should be increasing the awareness of the locality of the region through three key elements: (1) encouraging the local character, (2) enhancing the quality of the public realm, and (3) supporting adaptability and diversity.
- The third recommendation is to incorporate sustainable urban design strategies that invest in people. The idea is to shift the focus away from the physical aspects of design and toward the social systems of design. Sustainable initiatives often are the result of people who put effort into improving the quality of their living environments. Therefore, by increasing residents' involvement in the planning and design processes, suitability measures are more visible and tangible in the urban context. There are three steps that can contribute to investment in the human aspect of planning in design: economic security, ecological integrity, empowerment and responsibility. Design and development of neighborhoods in Phoenix should be based on these characteristics.

Conclusion

The overall picture of the Phoenix metropolitan area is one that has been shaped by national trends during the post-World War II era, by local conditions, and by regional policies. This resulted in a horizontal collection of auto-oriented central city-suburbs. As a result, any attempt of achieving urban sustainability in the region has been deterred by a fragmented regional government. In searching for a more sustainable model for the region it becomes apparent that the Dutch strategies and approaches represent a reliable source of knowledge for Phoenix because they have evolved dynamic and proactive ways to deal with urban pressures, yielding to more positive outcomes. Three main characteristics define the Dutch model (1) collaboration among the involved parties; (2) proactive solution to urban issues; and (3) planning policies that reflect current trends. However, the concepts, approaches, and strategies presented through a series of case studies in this research are only the first step in the dialogue about creating a more sustainable urban environment for Phoenix and its metropolitan region.

Reference List

Beatley T (2000) Green urbanism: learning from European cities. Island Press, Washington, DC

Bueren EM Boonstra C (2001) Sustainable design in the Netherlands. In: Edwards B (ed) Green Architecture, Special Issue Architectural Digest 71:4. Wiley and Sons, London, pp 76–81

Corbett J, Corbett M N (2000) Designing sustainable communities: learning from village homes. Island Press, Washington, D.C.

Coupland A (1997) Reclaiming the city: mixed use development. E & FN Spon, London

Downs A (2001) How the Phoenix region might cope with rapid growth. Maricopa Association of Governments, Phoenix

Lincoln Institute of Land Policy (2003) Making sense of place: Phoenix, the urban desert [videorecording]. Lincoln Institute of Land Policy, Cambridge, MA

Mclennan J (2004) The philosophy of sustainable design. ECOtone, Kansas City, Missouri

Ministry of Housing, P P a E (2000) Compact cities and open landscape: spatial planning in the Netherlands. The Hague, Unpublished manuscript

Maricopa Association of Governments (MAG) (2005) Regional report: a resource for policy makers in the Maricopa region. MAG, Phoenix

Morrison Institute for Public Policy (2000a) Hits and misses: fast growth in metropolitan Phoenix. Arizona State University, Tempe

Morrison Institute for Public Policy (2000b) The new economy: policy choices for Arizona. Morrison Institute for Public Policy, Arizona State University, Tempe

Morrison Institute for Public Policy (2006) Arizonans' attitudes toward science, technology, and their effects on the economy. Morrison Institute for Public Policy, Arizona State University, Tempe

Vlassenrood L (2004) Hybrid landscapes designing for sprawl in the Netherlands 1980–2004. BLAUWE KAMERS

Toward Dynamic Real-Time Informative Warning Systems

Robert Pahle and Filiz Ozel

Abstract A great deal of research has been done to improve safety in the built environment. Disaster preparedness plans have been developed for a variety of situations; however, these plans often fail to convey sufficient information flow to the target audience. The literature discusses a variety of arguments that both support and limit giving information to the target audience. One of the arguments for limiting information is the assumption that people could panic and increase the difficulties at hand. Yet, a literature search for this paper indicates that target audience panics only in very rare situations. Furthermore, it shows that most people seek information about the event and that they actually try to make good decisions to ensure their survival. A system that could bring information to persons during an emergency could therefore address a variety of decision-making challenges in these tense situations. The handling of information could encompass a variety of scales: from national to urban to building scale. This paper looks into the realm of dynamic, real-time informative warning systems that deliver information to occupants of buildings during emergencies. It builds on a cognitive model for navigation in the built environment. A theoretical approach for the design of the warning message is discussed. Since this is an ongoing research project, only the results of the initial assessment will be presented here.

Introduction

Emergency responders stand on the frontline when they help people out of dangerous situations. In these situations it is very important that they can rely on accurate information delivered in a timely manner. A lot of research has been done in this area resulting in the development of new technologies, some of which are now

R. Pahle(✉)

College of Design, Arizona State University, PO Box 871905, Tempe, AZ 85287-1905 USA,
E-mail: robert.pahle@asu.edu

commercially available. Today, emergency responders use technologies such as Geographic Information Systems (GIS), Global Positioning System (GPS), thermal cameras, and new communication technologies.

At the same time research shows that people in dangerous situations, for instance inside a building during a fire, ignore fire alarms, lose time by gathering belongings before leaving, or go toward the fire instead of away (Bryan 1981; Ozel 1993). Emergency situations inside buildings involve fast changing conditions that require precise and current information so that a person can understand the situation fully and can take proper action (Tong and Canter 1985a). The ultimate goal of the research summarized in this article is to study the methods of providing such information to occupants so that they can make better decisions and take faster action during emergencies.

The most common way to alert people to an emergency situation is a fire alarm system. There are two aspects of the alarm system that are especially important – the clarity and believability of the alarm (Tong and Canter 1985a). Since people tend to ignore fire alarm systems as an information source during a fire, many studies have been undertaken to identify other methods of conveying information in an emergency situation. Research has shown that improvement in the effectiveness of an alarm system can be achieved if it is combined with a graphical/aural explanation of the event (Bellamy 1990; Bellamy et al. 1988; Pigott 1989; Tong and Canter 1985a; Tong and Canter 1985b). For example, the use of public address systems and prerecorded messages was studied by Klein (1996), who found that it is crucial that the correct information be broadcasted so that the occupants are directed to safety and not inadvertently towards the fire. In addition to sound there are also prototypes of systems that display visual information about emergencies like BRESENS (Bellamy 1990; Pigott 1986). BRESENS was developed by the building research establishment (BRE). It is a small box that can display text on a 16 character LCD display. It can use a variety of sensors (e.g., smoke or heat) to detect a fire. Messages like "Ground Floor Fire Evacuate Now" can automatically be displayed (Bellamy 1990).

On the other hand, an informative warning system (IWS) can improve decision making during emergencies by presenting relevant real-time information to people in danger. Researchers like Bellamy (1990), Geyer (1988), Tong (1985), and Canter (1990) investigated different aspects of informative warning systems to measure information acquisition and response times among different modes of communication such as 2D/3D representation, audible alarms, and voice alarms (Bellamy 1990; Geyer et al. 1988), or to investigate principal uses of informative warning systems (Tong and Canter 1985b). Messages were usually displayed in a graphic representation either as text or as a drawing of a building. Further research regarding what information is actually needed during a particular phase of an emergency is needed.

Tong and Canter (1985b) investigated the perceived advantages and disadvantages of IWS. Only 27% of the population stated that an IWS would not be necessary, 42% would like to have such a system, 16% said it is useful, but not necessary, 9% found it necessary in some type of buildings and 6% were unsure. 70% of the responders said that the biggest advantage is the aid in the choice of escape route.

Concerns included that people might ignore the warning if the fire was not in their area, that it is expensive to implement, and that it might create a false sense of security (people don't understand the speed of fire spread, and the danger of failure). Other concerns were that it might cause a panic. These results support the importance of a safe evacuation route as well as the believability, reliability and validity of the information delivered.

Janis and Mann (1979) investigated decision making under stress. One of their main findings is that while an emergency progresses occupants will be under more and more time pressure resulting for instance in subjective and sub-optimal decisions. In other words, under time pressure, and as escaping gets more and more difficult, people cannot effectively process all the information they get. An IWS could take the current situation of the occupant into account and provide the appropriate amount of information. This amount of information might be called the "minimum essential information" necessary to make the decision that is most relevant to the current emergency situation. The information can be presented in visual and audible format, but to limit the scope of this paper we have concentrated on visual and cognitive aspects of an IWS.

In summary, this study poses two research questions: 1) exactly what information must be conveyed to the occupants of a building at different phases of a possible fire emergency, and 2) what are the most effective ways to display the most relevant information related to the layout of the building, the nature and location of the fire, the location of exits, etc.; and how can all of this information be best represented within the context of the current location of the individual occupant. Obviously, the spatial nature of this information needs to be addressed as well.

Different types of buildings and occupancies have different characteristics that must be incorporated into an informative warning system (IWS). For example, an office building and its occupants are very different from a hospital building and its occupants. Due to the complexity of circulation, building layout, and the unique emergency exiting problems that are posed by library buildings, a university library on the campus of Arizona State University in Tempe, Arizona was selected as a case to measure the methods with which an IWS system can be developed. In this ongoing research projects two survey studies have been conducted where patrons of Hayden Library at Arizona State University were interviewed and the design of preliminary IWS tools for this library was evaluated. Details of these studies will be discussed later in this article.

Background

Evacuation Model

An evacuation model tries to explain why people in an emergency situation behave as they do and to predict how they will behave in a future egress situation. There are many models that have been developed in the past.

Early models used the psychological stimulus-response (SR) approach that views evacuation as an instinctive primitive "flight" response to a threatening stimulus (Tong and Canter 1985a). Those theories evolved into "ball-bearing" models that are often referred to as environmental determinism, that treat individuals as unthinking objects that automatically respond to external stimuli (Sime 1992a; Sime 1992b). These models use complex mathematical equations to predict human movement (Pauls 1996) and use mainly carrying capacity of the structure and its various components for its predictions (e.g., Helbing et al. 2001).

While those models are good in predicting some aspects of an evacuation, they do not take into account the fact that behavior is influenced by personal evaluations of the socio-physical environment (Tong and Canter 1985a).

Since those models cannot give a full account of the behavior during egress, Sime developed an "affiliation model." He used eye-witness statements to support rules of affiliation, where people seek security with the familiar and move toward familiar places and people (Sime 1982; Sime 1983).

Later, Canter, Breaux, and Sime (1990) revised a general role/rule-based decision-making model for human behavior in fires (Figure 1). As one can see, information has a very important position in their model. They state that "People are known to actively seek information during an emergency in an attempt to clarify what is usually an ambiguous situation…" (Tong and Canter 1985a). Other authors (Annett 1969; Wood 1990) also recognize the importance of information in acting appropriately in the case of an emergency.

Recently Gwynne, Galea, Owen, and Lawrence (Gwynne et al. 2002) came up with a more holistic model to explain human behavior during emergencies inside buildings. Their model is composed out of four major factors:

1. Configuration of the enclosure: encompassing the effects of the geography of the structure, including exit widths, arrangements of exits, etc.
2. Procedures implemented within the enclosure: this would entail the familiarity of the occupants with the enclosure, especially exits, the training and activities of staff, location of alarms, and signage.
3. Environmental factors inside the structure. This describes the effects of heat toxins and irritant gases and smoke on the occupant's ability to navigate and make decisions.

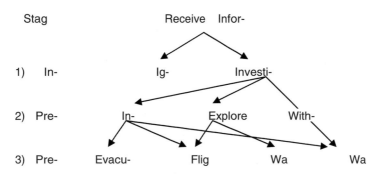

Fig. 1 General model of human behavior in fires (Canter et al. 1990)

4. Behavior of the occupants. This describes the culmination of all influences, incorporating group/social affiliation, the adoption of specific roles, the response to the individual to the emergency, likely travel speeds, and the ability to carry out desired actions (Gwynne et al. 2002).

Although the model characterizes major factors that influence exit behavior, it does not fully acknowledge the importance of correct real-time information during the emergency.

The approach taken in the study summarized in this article is to use real-time information about the emergency, the building, and other occupants to improve the decision-making process of occupants. This information can not be static or predetermined, since—as discussed above—the conditions of an emergency are fast changing, and outdated information could lead occupants to unsafe locations (e.g., Klein 1996). To stress the importance of real-time information during emergency egress, the model by Gwynne can be supplemented with a new factor "real-time information." The factor "real time information" comprises all information that an occupant is able to gather and process at a specific time during an emergency and that can potentially have an impact upon the occupant's behavior. The model could therefore be revised as can be seen in Figure 2.

Information Seeking Process

Fire-safety researchers have identified the nature of human behavior during fires as episodic. (Canter et al. 1990; Ozel 1993) Each episode is characterized by a specific goal and therefore by a specific behavior, such as to investigate the situation, alert

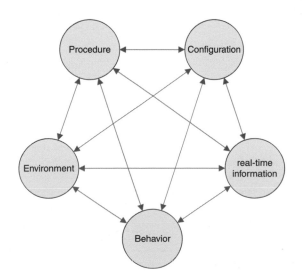

Fig. 2 Modified model of the factors influencing evacuation after Gwynne et al. (2002)

others, fight the fire, exit the building or protect valuables. Once one episode is finished a new episode starts, for example, investigating of the situation is followed by protecting valuables, which is followed by exiting the building (Canter et al. 1990). When the percentages for all episodes are considered together, the percentage of people who investigate is as high as 56% during a fire emergency (Bryan 1981; Ozel 1993). In this regard, investigation means to seek information for decision making.

In order to understand the exact information that might be needed for display in an IWS, research done by Canter, Breaux, and Sime can provide the preliminary lead (Canter et al. 1990). They investigated different fires and created a general model of what people do during a fire emergency. For each action they found it is possible to identify the information needed to support the action. The researchers defined those actions based on corresponding information needs. Table 1 gives the overview.

Table 1 Actions during emergencies (after Bryan 1981; Ozel 1993)

Action (*)	Description (**)	Needed Information (**)
Investigate	gather a general understanding of what exactly happened after getting a clue or receiving a warning (in their model it seems that everybody investigated)	type of emergency, location of emergency, location of oneself, building layout
Investigate source	see where the fire is and if it can be extinguished	location of fire, distribution of fire, size of fire, maybe speed of spread, maybe toxicity/heat, maybe distribution/spread of smoke, location of emergency equipment, building layout
Warn/instruct others	help others to understand what happened and what to do now	type of emergency, location of the fire, location of exits, maybe safe route out of the building, location of other people, estimated time left
Evacuate	leave the building	location of oneself, location of exits, safe route out of the building, building layout
Evasive	leave the dangerous area	location of oneself, location of fire / smoke safe / refuge / dangerous areas, safe route away from the fire
Wait for help	stay in the room (after returning from investigation) and wait for emergency responders to arrive	time to arrival of emergency responders, updates on the general situation
Fight the fire	try to extinguish the fire	location of firefighting equipment, size of fire, location of fire toxicity/heat distribution / spread of smoke

(*) Actions as defined by Bryan (1981).
(**) Descriptions and Information needs defined by the researchers of this study.

The information needs can be summarized as follows:

- type of emergency
- location / distribution / speed of spread of emergency
- building layout
- location of oneself
- location of exits
- location of other people (e.g., relatives)
- safe route out of the building
- location of emergency equipment
- safe / refuge / dangerous areas
- time until arrival of emergency responders
- estimated time left

Humans as Information Processing Entities/Decision Makers

If the information that the occupants seek were to be available (e.g., on a monitor) the question still remains: How do they process, understand and act upon the information in this time-critical and stressful environment?

Filiz Ozel wrote a summarizing paper on this topic with a focus on the built environment (Ozel 2001). The topics of this paper will be discussed with regard to real-time informative warning systems.

The first aspect of how humans process information is how they cope with the information at hand. Three different behaviors can be postulated   acceleration, avoidance, and filtration (Miller 1960).

1. Acceleration is the processing of information at a faster rate. It might cause errors due to temporary overload of memory or the processing capacity.
2. Decision avoidance can result in random choices or in choices of momentarily striking characteristics in a set of choice alternatives.
3. Filtration is the compromise of choosing only the subjectively important data for consideration.

Miller's insight leads to the understanding that an informative warning system has to take care not to overload the decision maker (i.e., occupant/user of the IWS) in a time-critical situation. G. Miller gives an indication of a manageable amount of information (Miller 1956). He states that up to seven (+/–2) different elements might be distinguished in a class of elements, but those classes might be combinable.

Avoidance and filtration can be reduced by objectively presenting more important data for a decision so that the observer's attention can be better focused on it. This might include a dynamic ranking on possible choice alternatives based on objective factors like proximity of the occupant to fire and stairs.

Visual Presentation of the Warning

To examine how information in an IWS should be presented, it is important to understand the type of information it will process. The examination of information needs of occupants earlier in this paper revealed that a lot of the needed information has a spatial component. For instance, location/distribution of emergency, location of oneself, location of exits, location of other occupants, a safe route out of the building, location of emergency equipment, and safe/refuge/dangerous areas. Only the type of emergency, the time of arrival of emergency responders, and the estimated time until a location becomes dangerous are not spatial. Therefore the visual appearance of the warning system should be spatial. This is supported by Bellamy's (1990) research.

Fire researchers also looked at the behavior of occupants that are familiar with a particular building. In general, people (while evacuating) tend to use routes they use every day or that they are familiar with compared to the shortest route to a safe location. Tests with mockup models of IWS showed that familiarity significantly affected the number of subjects that would interpret the warning as genuine fire alarm (67% of the familiar population vs. 51% of the unfamiliar population) (Bellamy 1990). A remaining question here is how people will react to route suggestions of an IWS or if a route should even be suggested.

Another aspect that needs consideration is time. As many researchers point out, emergencies, especially fire emergencies, are fast changing in nature (Tong and Canter 1985a). At one point in time a fire is confined to a particular area, but moments later it might have spread, blocking exits that were available before. Warning systems that do not incorporate the dynamics of emergencies may produce fatal results. For instance, the alarm system at the Düsseldorf airport fire used pre-recorded messages, sending people toward the fire instead of to a safe location (Klein 1996). Therefore an IWS has to be dynamic and should show an updated version of the warning as soon as new information becomes available.

Dimensions and Limitations of a Dynamic, Real-Time Informative Warning System

Warning systems could be applied in a variety of scales. It can operate on a national scale, on a city or neighborhood scale, or on a building scale. For this research the focus is on building scale. Another dimension of a warning system can be the devices the information is distributed to. The range goes from mobile devices to stationary desktops and televisions.

Depending on the choice of the information dissemination devices, different models can be assigned to the navigational needs of the user. A mobile device like a cell phone or a pocket-pc allows for a more perceptual conceptualization of navigation (sometimes also referred to as blind navigation). A message on the display will say something like "follow the arrow that is displayed on the device" and a big arrow is displayed in the direction the occupant should go.

In case of a stationary desktop a more cognitive model applies. The occupant has to have a mental model of the building and a proper reference (enough physically distinct features) to locate and orient themselves in space and find appropriate target locations (like exits or rescue areas). An additional mental effort has to be made to revise a safe route toward the target (Passini 1992). Despite a more challenging mental effort for the second model there are some advantages. For instance, many occupants would like to know that people important to them are safe. In the cognitive model it is possible to display information on where other people are in a straight forward manner, since the whole building is shown. It may also help if later during the exit a route is blocked, since a more complete mental model of the building and the dangerous areas is available.

Additional assumptions are that an IWS has to be specifically designed for a building and a user group. As it will be shown later in this paper, every building type, e.g., the library that will be discussed, can have very specific elements that can be used to complete the mental model of the space. In the case of the library these are the call numbers that are available on almost every bookshelf. And also each user group has different needs. Where the staff will know the building layout rather well, library patrons may be unfamiliar with the spatial organization. They may also not understand the location of stairs in areas with no public access.

Research Study at Arizona State University

ASU's Hayden Library building was chosen to study the specifics of an IWS tool according to the patrons of the library. The decision to choose this library building as part of this case study was done not only because library buildings pose unusual challenges for its patrons, but also because Hayden Library's building is a complex structure with an underground entrance and a connection to the actual building through the main lobby area. Looking at the functions in this building and its occupancy type led to some assumptions about the users:

Age	18–60
Familiarity	familiar-unfamiliar
Disability	not disabled
Education	educated
Role in the building	Library patron or staff
Study habit	study alone

An initial study to understand the nature of a possible IWS system's graphic representation was originally done in the summer of 2006. This was planned as a pilot study of a very small number of students using an interview technique. For the pilot study, a preliminary graphic design for a warning system was created and

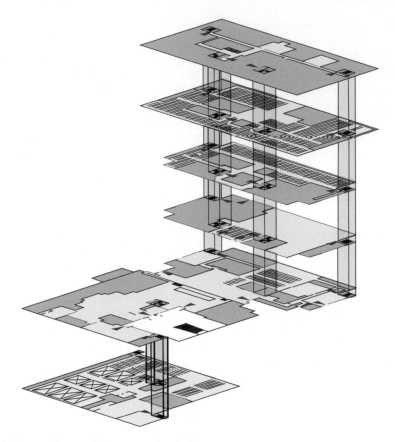

Fig. 3 Layout of the preliminary warning design

tested. After having only four participants complete the survey, which included the image shown in Figure 3 to indicate the spatial characteristics of the library, it became clear that there was a need to approach the design of the graphics more systematically.

The initial design was based on the architect's floor plans with some common sense principles and a basic architectural understanding of how to render a building in 3D. Several issues were raised by the participants regarding this graphic, including the following:

- the test persons could not correctly identify where they were
- red arrows were not interpreted as exit doors
- the lines to signify the stairs were not recognized
- lines in between layers were misinterpreted as elevators

The only insight gained from this study was the idea to use call numbers located on the bookshelves as a directional marker (or landmark). This feature was eventually used in the next version of the IWS graphics.

After a further literature search was conducted the principles of design outlined in *Universal Principles of Design* (Lidwell et al. 2003) were assessed to be applicable to the design of the graphic interface for the IWS system. Because the focus of this book is information delivery through graphic design, the principles listed are applicable to the design of IWS systems. The authors describe proven standard design principles that could be used to optimize information delivery between display and the user. In reviewing these design principles, 44 were found to be applicable to the task at hand. Those rules where then matched against the theories of time pressure and stress as well as wayfinding. Each principle was then categorized based on its importance: very important (++), important (+), neither important nor unimportant (0), unimportant (-) or irrelevant (--).

Each design principle was then put into one of four groups. The groups describe the aspect of the IWS that should be improved. Finally the principles were ranked based on the scores for the theories. This allows reducing the amount of design principles to use for each group of improvements to between two and six. Two combined rankings were given: 1) most important and 2) important (see Table 2;

Table 2 Ranked design principles

	Design Principle	Description	Time Pressure and Stress	Wayfinding
Improve understanding				
2	Iconic representation	The use of pictorial images to improve the recognition recall of signs and controls	+ faster to understand	+ universally understand-able
1	Performance Load	The greater the effort to accomplish a task, the less likely the task will be accomplished successfully	++ it is very important to simplify to make sure in the given time the most important facts are understood	+ For wayfinding it would be better to have a system that is very easy to understand to optimize performance
2	Ockham's Razor	Given the choice between functionality equivalent designs, the simplest design should be selected	+ important	+ important
1	Chunking	A technique of combining many units of information into a limited number of units or chunks, so that the information is easier to process and remember	++ especially important to simplify the understanding of information in time-critical environments	+ widely used technique to group information on navigational signs in chunks of three

(Continued)

Table 2 (continued)

	Design Principle	Description	Time Pressure and Stress	Wayfinding
2	Flexibility-Usability Tradeoff	As the flexibility of a system increases, its usability decreases	+ important in terms of what decisions the system should support.	+ important in terms of what decisions the system should support.
1	Interference Effect	A phenomenon in which mental processing is made slower and less accurate by competing mental processes	++ it is very important to make sure all signs and layouts are clearly understood and the information used is used the same way in our daily life	+ it is imortant to be clear about meaning of signs

Help IWS users to make the right decision

	Design Principle	Description	Time Pressure and Stress	Wayfinding
1	Operant Conditioning	A technique used to modify behavior by reinforcing desired behaviors, and ignoring or punishing undesired behaviors	++ it is very important to educate (and drill) people to behave correctly and timely in an emergency	+ it is of great value to have people understand signs correctly and take proper action
1	Mental Model	People understand and interact with systems and environments based on mental representations developed from experience	+ it is important to avoid working against common mental models to avoid a time lag to action	++ it is very important to build mental models of the spaces (building) so that also difficult wayfinding situations during emergency egress can be mastered
2	Hierarchy	Hierarchical organization is the simplest structure for visualizing and understanding complexity	++ very important to improve the decision process	+ for a visual representation to support wayfinding a visual hierarchy of paths is important

(Continued)

Table 2 (continued)

	Design Principle	Description	Time Pressure and Stress	Wayfinding
2	Constraint	A method of limiting actions that can be performed on a system	+ it is important under time pressure and stress to show only the options that are possible or valuable in a specific situation	+ it is important to understand the tradeoff between flexibility of route choice and other tradeoffs like crowd management
2	Hick's Law	The time it takes to make a decision increases as the number of alternatives increases	++ it is very important to provide the minimal necessary number of alternatives to make sure the available time is used right	+ it is important to provide the minimal necessary number of alternatives
1	Highlighting	A technique for bringing attention to an area of text or image	++ it is very important to guide the user to optimal solutions in the least amount of time. ++ Subjective selection can be focused ++ It can relieve stress by supporting decision making	++ it is important to support the generation of optimal decision plans

Strategies to better remember information to improve decision making during emergencies

	Design Principle	Description	Time Pressure and Stress	Wayfinding
1	Picture Superiority Effect	Pictures are remembered better than words	++ This effect can be used to speed up recognition of locations and also recognition of threads	++ For wayfinding the IWS could be enriched with pictures to recognize locations and reduce errors in wayfinding

(*Continued*)

Table 2 (continued)

	Design Principle	Description	Time Pressure and Stress	Wayfinding
2	Serial Positions Effect	A phenomenon where items presented at the top or the end of a list are more likely to be recalled than items in the middle of the list	+ it can be important that when presented with choices in an IWS, the most important ones are at the top and the bottom of the list	- not as important

Help to make an IWS better readable

	Design Principle	Description	Time Pressure and Stress	Wayfinding
2	Three-Dimensional Projection	A tendency to see objects and patterns as three dimensional when certain visual cues are present (Interposition, Size, Elevation, Linear Perspective, Texture Gradient, Shading, Atmospheric Perspective)	+ three dimensional projection can improve decision making under timepressure and stress	+ three dimensional projection can improve spatial orientation and therefore wayfinding
1	Visibility	The usability of a system is improved when its status and methods of use are clearly visible	++ very important to ensure function of an IWS under stress conditions	++ very important to ensure all information that is needed for the wayfinding task is clearly visible
2	Similarity	Elements that are similar are perceived to be more related than elements that are dissimilar	++ very important to make sure that all elements that are related (like exits) are similar to ensure everything is understood correctly	++ very important to make sure that all elements that are related (like exits) are similar to ensure everything is understood correctly

(Continued)

Table 2 (continued)

	Design Principle	Description	Time Pressure and Stress	Wayfinding
1	Uniform Connectedness	Elements that are connected by uniform visual properties, such as color, are perceived to be more related than elements that are not connected	++ very important to make sure that all elements that are related (like exits) are similar to ensure everything is understood correctly	++ very important to make sure that all elements that are related (like exits) are similar to ensure everything is understood correctly
2	Entry Point	The point of physical or attentional entry into a design	++ it is very important that the people are focused on the most important facts first, this can save valuable time	+ it is important to build an IWS so that the wayfinding decision plans can be deduced in a logical successive manner
1	Signal to Noise Ratio	The ratio of relevant to irrelevant information in a display. The highest signal to noise ratio is desirable in a design	++ very important to focus on the most important information and get rid of everything that is not needed for the task. This improves understandability and speeds up the process of reading the IWS	+ this improves the understandability and reduces the errors of the wayfinding task

only entries with rankings 1 or 2 are shown). The table should be read as follows: If one were to improve the understandability of a warning message, the first choices to work on would be "performance load," "chunking," and "interference effect." Secondary options would be "iconic representation," "Ockham's Razor," and "Flexibility–Usability Tradeoff." It has to be noted that there is no such thing as the perfect design. It highly depends on the factors described above and many of them are counteracting.

After identifying these design principles they were used to generate a second version of the IWS graphic interface (Figure 4). This design was tested on about 20 persons. The population was sampled at different times of the day and at different locations throughout the building to reduce the bias based on time of the day and the floor of the building. The results are in strong contrast to the preliminary test.

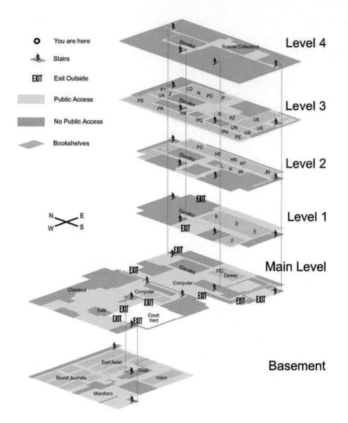

Fig. 4 Revised version of the information design

The first question that was asked is to identify one's own location (please observe that it is not given in Figure 4). All participants were able to identify their current location. This may be attributed to several changes in the map that were based on the ranked design principles identified earlier.

- display of proper floor labels
- de-cluttering (polygons are not bordered by black lines anymore, glass surfaces are not shown anymore)
- the introduction of the library call numbers (letter codes on the floor plans) seems to be really helpful to the users to orient themselves in the almost symmetrical floor-plan of the upper stories, one person even placed a point on one side, checked the call number posted on the wall and revised his own location on the map

The occupants were also able to identify the exits. This is probably based on the EXIT symbol that was placed at the exit locations. The exit symbol was modeled exactly after the original exit signage throughout buildings; therefore it can be understood by illiterate people who have learned the meaning of the symbol.

Not all occupants however were able to orient themselves properly inside the building. Some were actually trying to go into the wrong direction (e.g., south instead of north).

Outlook

As a summary, the study aimed at investigating the following research questions: 1) exactly what information must be conveyed to the occupants of a building at different phases of a possible fire emergency in a building, and 2) what are the most effective ways to display the most relevant information related to the layout of the building, the nature and location of the fire, the location of exits, etc.

In the pilot study as well as in the more extensive study, the researchers have tried to address both issues, but more so the second one. The focus has been primarily on the graphic design of the IWS system, and whether the graphics and text used were interpreted correctly by the users. Also, the nature of the graphics was mostly based on the information needs of occupants for exiting purposes with heavier emphasis on orientation and being able to spatially function in the building. Proper spatial orientation is obviously essential for most actions people would want to take during an emergency in a building. Therefore, the initial focus of both studies has been on the ability of the IWS system to convey information that will help occupants spatially orient themselves. Tests regarding this will continue, and the graphics for the IWS system will be further refined in support of successful spatial orientation. The next step after this will be how to convey information regarding the progress of the emergency situation. The researchers have had a glimpse into this problem in the latest version of the IWS graphics interface, when they tried to represent areas that were not safe to go into. This aspect of the IWS graphic will be more closely scrutinized through additional studies undertaken with the participation of the occupants of the Hayden Library building at ASU.

Citations

Annett J (1969) Feedback and human behaviour. Penguin, Harmondsworth

Bellamy LJ, Geyer TA, Max-Lino R, Harrison PI, Bahrami Z, Modha B (1988) An evaluation of the effectiveness of the components of informative fire warning systems. In: Sime J (ed) Safety in the built environment. SPON, pp 36–47

Bryan JL (1981) Implications for codes and behavioral models from the analysis of behavior response patterns in fire situations as selected from the Project People and Project People II study programs. University of Maryland, Department of Fire Protection, College Park, MD

Canter DV, Breaux J, Sime Jonathan D (1990) Domestic, multiple occupancy, and hospital fires. In: Canter DV (ed) Fires and human behaviour. Fulton, London, UK

Geyer TA, Bellamy LJ, Max-Lino R, Harrison PI, Bahrami Z, Modha B (1988) An evaluation of the effectiveness of the components of informative fire warning systems. In: Sime J (ed) Safety in the built environment. SPON, pp 36–47

Gwynne S, Galea ER, Owen M, Lawrence PJ (2002) An investigation of the aspects of occupant behaviour required for evacuation modeling. In: DeCicco PR (ed) Evacuation from fires, vol 2. Baywood Publishing Company, INC., Amityville, New York

Helbing D, Farkás IJ, Molnár P, Vicsek T (2001) Simulation of pedestrian crowds in normal and evacuation situations. In: Schreckenberg M, Sharma SD (eds) Pedestrian and evacuation dynamics. Springer, Berlin, Gerhard-Mercator University in Duisburg, Germany, pp 21–58

Janis IL, Mann L (1979) Decision making: a psychological analysis of conflict, choice, and commitment. Free Press Collier Macmillan, New York, London

Klein RA (1996) The Dusseldorf airport fire. Fire Engineers Journal May 1996:18–23

Lidwell W, Holden K, Butler J (2003) Universal principles of design: 100 ways to enhance usability, influence perception, increase appeal, make better design decisions, and teach through design. Rockport, Gloucester, Mass.

Miller GA (1956) The magical number seven, plus or minus two: some limits on our capacity for processing information. Psychological Review 63:81–97

Miller JG (1960) Information input overload and psychopathology. American Journal of Psychiatry 116:695–704

Ozel F (1993) Computer Simulation of Behavior in Spaces. In: Marans RW, Stokols D (eds) Environmental simulation: research and policy issues. Plenum Press, New York

Ozel F (2001) Time pressure and stress as a factor during emergency egress. Safety Science 38:95–107

Passini R (1992) Wayfinding in architecture, Paperback edn. Van Nostrand Reinhold, New York

Pauls J (1996) Movement of People. In: Dilenno PJ et al. (eds) The SFPE handbook of fire protection engineering, 2nd edn. National Fire Protection Agency, Quincy, MA, pp 3–263, 263–285

Pigott BB (1986) The scope for intelligent fire detection systems. In: International fire safety and security exhibition and conference, pp 19–39

Pigott BB (1989) Fire detection and human behavior. In: Wakamatsu T, Hasemi Y, Seizawa A, Seeger P, Pagni P, Grant C (eds) Second international symposium on fire safety science, pp 573–581

Sime JD (1982) Group affiliative behaviour during flight to building exits. In: 7th international conferrence on people and their physical surroundings. University of Barcelona, Spain

Sime JD (1983) Affiliative behaviour during escaping to building exits. Journal of Environmental Psycology 3:21–42

Sime JD (1992a) Group affiliative behaviour during flight to building exits. In: Seventh international conference on people and their physical surroundings, Spain

Sime JD (1992b) Human behaviour in fires: summary report. Central Fire Brigades Advisory Council for England and Wales

Tong D, Canter DV (1985a) The decision to evacuate: A study of the motivations which contribute to evacuations in an event of a fire. Fire Safety Journal: 257–265

Tong D, Canter DV (1985b) Informative warnings: in situ evacuations of fire alarms. Fire Safety Journal: 267–279

Wood PG (1990) Survey of behaviour in fires. In: Canter (ed) DV Fires and human behaviour, 2nd ed. Fulton, London, UK, pp 83–95

Printing: Krips bv, Meppel, The Netherlands
Binding: Stürtz, Würzburg, Germany